AF718337

Student Study Guide

Geralyn M. Koeberlein, Mahomet-Seymour High School

Daniel C. Alexander, Parkland College

Elementary Geometry for College Students, Third Edition

Daniel C. Alexander, Parkland College

Geralyn M. Koeberlein, Mahomet-Seymour High School

HOUGHTON MIFFLIN COMPANY BOSTON NEW YORK

Senior Sponsoring Editor: Lynn Cox
Editorial Associate: Melissa Parkin
Senior Manufacturing Coordinator: Jane Spelman
Marketing Manager: Ben Rivera

Copyright © 2003 by Houghton Mifflin Company. All rights reserved.

No part of this work may be reproduced or transmitted in any form or by any means, electronic or mechanical, including photocopying and recording, or by any information storage or retrieval system without the prior written permission of Houghton Mifflin Company unless such copying is expressly permitted by federal copyright law. Address inquiries to College Permissions, Houghton Mifflin Company, 222 Berkeley Street, Boston, MA 02116-3764.

Printed in the U.S.A.

ISBN: 0-618-22178-6

4 5 6 7 8 9—MZ—07 06 05 04

Student Study Guide

The Student Study Guide for *Elementary Geometry for College Students, Third Edition* includes the following components:

*Includes solutions for odd-numbered problems from each section and solutions to all problems found in chapter reviews and appendices.

Student Study Guide

How to Study Geometry

Textbook:

1. Read the textbook word for word.
2. Pay special attention to undefined terms and definitions. The topics of the textbook that follow build upon this terminology.
3. When you encounter a definition:
   a. Make a drawing (to reinforce the concept).
   b. Think of other examples and counter-examples (things that do *not* meet the description) to compare with the defined term.
   c. State the definition in your own words (Do not just memorize a cluster of words).
4. When you encounter a postulate or theorem:
   a. Read and reread until you understand the statement.
   b. Make a drawing to illustrate the statement.
   c. Ask if the meaning would change if words were added or deleted.
   d. State the postulate or theorem in your own words.
   e. For a given theorem, ask yourself if the statement makes sense. Why should it be true?

> NOTE: The student should create an index card for each important term, postulate, or theorem. On the card, write the statement (the definition, postulate, or theorem), illustrate with a drawing, and include an example.

5. When you encounter a textbook example:
   a. Read it step for step, justifying each step as it unfolds.
   b. Refer to the drawings that are provided in arriving at each conclusion.
   c. Try repeating the steps of the example with the book closed.
6. When you encounter a completed proof (like an example):
   a. Note the ordering of "statements" and the justifying "reasons."
   b. Reason from the given information using the drawings that are provided.
   c. Consider the statements *in reverse order* noting that each statement is true because it follows logically from a statement that precedes it.
   d. Try repeating the steps with the book closed.
7. It is good to keep track of the many methods that achieve a particular goal. For example, write a list of methods for proving that *triangles are congruent* or that *lines are parallel.* For reference, see the list of postulates and theorems in the appendix of the textbook.

Assignments:

1. Be sure to complete as many of the assigned problems as possible. One cannot expect to solve problems found on quizzes or tests if one has not practiced by doing assignments beforehand.

> If you have a Student's Solutions Manual, you must use it wisely; try the problem first, and then use the manual it to help you with a difficult step. DO NOT just read the author's solution; in no way does that guarantee that you the student could generate that solution.

2. If a problem of the assignment seems difficult, try the next one. Unlike a cluster of very similar algebra problems, consecutive geometry problems are often very different. Due to a discovery made in solving a later problem, you can sometimes return to and solve the "difficult" problem.
3. It is important to work on an assignment solo at first. When having difficulty doing an assigned problem, you should look for a textbook example of a similar nature to serve as a guide.
4. When there are difficulties or you are frustrated, it is worthwhile to seek the help of a tutor or of a classmate. You may need to make an appointment to speak to the instructor.

Preparing for a Test:

1. Handwrite a list of the definitions, postulates, and theorems included in the test material. Study the list thoroughly and then try to write each statement in your own words.
2. Study the chapter summaries and attempt chapter review exercises found at the end of each chapter of the textbook.
3. Repeat the assigned problems, especially those that caused difficulty.
4. If questions remain, consult your instructor, a tutor, or a study partner.
5. Preparation for a test really begins with the first day of class. It may be very important for your success in a course like geometry to form a study group . . . to study with on a regular basis and before tests.

General Behavior:

1. Attend all classes. Arrive on time!
2. Develop an interest. Ask questions in class.
3. Find a study buddy or form a study group that meets regularly.
4. Utilize the instructor's office hours and tutoring services.

Section-by-Section Objectives

The following are the student goals for Elementary Geometry for College Students, Thrid Edition

Chapter One: Line and Angle Relationships

Section 1.1 : Statements and Reasoning

1. determine whether a collection of words/symbols forms a statement;
2. form the negation of a given statement;
3. form the conjunction, disjunction, or implication determined by two statements;
4. recognize the hypothesis/conclusion of a conditional statement;
5. state the three types of reasoning used in geometry;
6. determine the type of reasoning used in a specific situation; and
7. recognize/apply the Law of Detachment.

Section 1.2: Informal Geometry and Measurement

1. describe the terms point, line, and plane;
2. become familiar with geometric terms such as collinear, line segment, and angle;
3. measure a line segment with a ruler/measure an angle with a protractor;
4. write equations based upon statements involving midpoint, bisect, and congruent;
5. recognize the terms right angle, straight angle, and perpendicular;
6. use the compass to construct a line segment of specified length; and
7. use the compass to determine the midpoint of a given line segment.

Section 1.3: Early Definitions and Postulates

1. know the parts of a mathematical system: undefined terms, definitions, postulates, and theorems;
2. recognize the need for/characteristics of a precise definition;
3. know the definition/symbol for line segment and its length;
4. accept and state the initial postulates involving lines and planes (in your own words);
5. use the Segment-Addition Postulate to write equations;
6. understand the concepts parallel lines and parallel planes; and
7. recognize the significance of the term "unique" as it applies to geometry.

Section 1.4: Angles and Their Relationships

1. know the definition/symbol for angle and its measure;
2. understand/use terms related to angles (like sides, vertex, etc.);

3. state/apply postulates involving an angle(s);
4. recognize the type of angle shown/measured: acute, right, obtuse, and straight;
5. use the Angle-Addition Postulate to write equations;
6. know the classifications of pairs of angles: adjacent, congruent, complementary, supplementary, and vertical;
7. use the compass to construct an angle congruent to a given angle; and
8. use the compass to construct the bisector of a given angle.

Section 1.5: Introduction to Geometric Proof

1. demonstrate the two-column form of a proof;
2. understand the role of the Given, Prove, and Drawing for a proof problem;
3. provide reasons that justify statements supplied in partial proofs; and
4. provide statements that are justified by the reasons supplied in partial proofs;

Section 1.6: Relationships: Perpendicular Lines

1. know/apply the definition of perpendicular lines in practice and proof;
2. develop an understanding of the concept of "relation;"
3. understand/apply the reflexive, symmetric, and transitive properties of congruence;
4. construct the unique line perpendicular to a line at a point on the line; and
5. construct the unique perpendicular-bisector of a given line segment.

Section 1.7: The Formal Proof of a Theorem

1. state the hypothesis and conclusion of a given theorem;
2. state the five written parts of the formal proof of a theorem;
3. make the "Drawing" for the proof of a theorem based upon the hypothesis of the theorem;
4. write the "Given" for the proof of a theorem based upon its hypothesis and the "Drawing";
5. write the "Prove" for the proof of a theorem based upon its conclusion and the "Drawing";
6. state/apply theorems involving perpendicular lines, complementary angles, and so on; and
7. construct/complete the formal proof of a theorem.

A Look Beyond Chapter 1: Historical Sketch of Euclid

1. recognize Euclid (Greek) as a principal contributor to geometry; and
2. know that the Elements of Euclid are the basis for any plane geometry textbook.

Chapter Two: Parallel Lines

Section 2.1: The Parallel Postulate and Special Angles

1. construct the perpendicular line from a point not on a given line to that line;
2. recognize when two lines, a line and a plane, or two planes are perpendicular;
3. recognize when two lines, a line and a plane, or two planes are parallel;
4. define parallel lines and parallel planes;
5. understand/apply terms such as transversal, corresponding angles, etc.;
6. state/apply initial postulates involving parallel lines; and
7. state/complete/apply selected theorems involving given parallel lines.

Section 2.2: Indirect Proof

1. know the true/false relationships between a conditional statement and its converse, inverse, and contrapositive;
2. state/apply the Law of Negative Inference;
3. state/apply the method of indirect proof; and
4. recognize that negations/uniqueness theorems are often proved indirectly.

Section 2.3: Proving Lines Parallel

1. state/apply/prove selected theorems establishing that lines are parallel; and
2. construct the line parallel to a given line through a point outside the line.

Section 2.4: The Angles of a Triangle

1. know definitions of triangle and related terms (vertices, sides, etc.);
2. classify triangles by their sides (scalene, isosceles, equilateral);
3. classify triangles by their angles (acute, right, obtuse, and equiangular);
4. know/apply the theorem, "The sum of angles of a triangle is 180°."; and
5. state/apply/prove the corollaries of the theorem stated in (4).

Section 2.5: Convex Polygons

1. know the definitions of polygon and related terms;
2. classify polygons as convex/concave and by their numbers of sides;
3. determine the number of diagonals for a polygon of n sides;
4. state/apply theorems involving sums of angle measures of a polygon;
5. classify polygons as equiangular/equilateral/regular; and
6. recognize a figure that is a polygram/regular polygram.

A Look Beyond Chapter Two: Non-Euclidean Geometries

1. know that the Parallel Postulate characterizes Euclidean (plane) Geometry;
2. recognize the existence of geometries other than plane geometry.

Chapter Three: Triangles

Section 3.1: Congruent Triangles

1. state the definition of congruent triangles;
2. determine the correspondences between parts of congruent triangles;
3. determine the included side (angle) for 2 angles (2 sides) of a triangle; and
4. know/apply these methods for proving congruence of triangles: SSS, SAS, ASA, and AAS.

Section 3.2: Corresponding Parts of Congruent Triangles

1. use CPCTC to symbolize "Corresponding parts of congruent triangles are congruent";
2. recognize the types of conclusions that can be established by using CPCTC;
3. use markings on congruent triangles to indicate their corresponding parts;
4. state/apply the HL theorem;
5. state/apply the Pythagorean Theorem; and
6. determine the method that establishes that triangles are congruent.

Section 3.3: Isosceles Triangles

1. distinguish between angle-bisector of angle of triangle, altitude of triangle, perpendicular-bisector of a side of triangle, and median of a triangle;
2. know that a triangle has three angle-bisectors/altitudes/medians/perpendicular-bisectors of sides;
3. decide whether an auxiliary line is determined/overdetermined/underdetermined;
4 state/use "If two sides of a triangle are congruent, the angles opposite these sides are congruent.";
5 state/apply "If two angles of a triangle are congruent, the sides opposite these angles are congruent."; and
6. state/apply the definition of perimeter of a triangle.

Section 3.4: Basic Constructions Justified

1. construct/justify the construction of an angle congruent to a given angle;
2. construct/justify the angle-bisection method;
3. construct/validate the construction of line segments of specified length;
4. construct/validate the construction of angles of specified measure; and
5. construct/validate the construction of selected regular polygons.

Section 3.5: Inequalities in a Triangle

1. know/use the definition of "is less than";
2. know the relationships found in the lemma (helping theorems) of this section;
3. state/apply theorems involving inequalities in a triangle;
4. state/apply corollaries involving the length of a line segment from a point not on a line (or plane) perpendicular to the line or plane; and
5. state/apply the Triangle Inequality (or its alternate form).

A Look Beyond Chapter 3: Historical Sketch of Archimedes

1. know that Archimedes' contributions to geometry included a good approximation of π; and
2. know that Archimedes was famous for inventions such as the catapult.

Chapter Four: Quadrilaterals

Section 4.1: Properties of a Parallelogram

1. state definitions for quadrilateral and parallelogram;
2. state/apply/prove selected theorems involving given parallelograms;
3. use angle measures of a parallelogram to determine its longer/shorter diagonal; and
4. determine speed/direction of an airplane whose motion is subject to the wind.

Section 4.2: The Parallelogram and Kite

1. recognize that the parallelogram/kite each have two pairs of congruent sides;
2. know that quadrilaterals with congruent opposite sides are parallelograms;
3. know that quadrilaterals with a pair of congruent and parallel sides are parallelograms;
4. know that quadrilaterals with diagonals that bisect each other are parallelograms;
5. know that the kite has one pair of opposite angles that are congruent;
6. know that the kite has a diagonal that is the perpendicular-bisector of the other diagonal; and
7. state/apply the theorem in which the midpoints of two sides of a triangle are joined.

Section 4.3: The Rectangle, Square, and Rhombus

1. state definitions for the rectangle, square, and rhombus;
2. state/apply/prove theorems involving the rectangle/square/rhombus;
3. state/apply/prove corollaries involving the rectangle/square/rhombus; and
4. apply the Pythagorean Theorem to quadrilaterals..

Section 4.4: The Trapezoid

1. know the terminology related to the trapezoid and isosceles trapezoid;
2. state/apply/prove selected theorems and corollaries involving trapezoids; and
3. state/apply "If three (or more) parallel lines intercept congruent segments on one transversal, then they intercept congruent segments on any transversal."

A Look Beyond Chapter 4: Historical Sketch of Thales

1. know that Thales was called the "Father of Geometry"; and
2. know that Thales was famous for the wisdom he displayed in everyday affairs

Chapter Five: Similar Triangles

Section 5.1: Ratios, Rates, and Proportions

1. state/apply the terms ratio, rate, and proportion;
2. know the terminology (means, geometric mean, etc.) related to proportions;
3. state/apply the Means-Extremes Property (of a proportion); and
4. understand/apply further properties of proportions.

Section 5.2: Similar Polygons and Triangles

1. form an intuitive understanding of the concept "similarity of figures";
2. determine the correspondences between the parts of similar polygons;
3. state/apply the definition of similar polygons;
4. state/apply the AA corollary;
5. recognize/apply CSSTP, meaning "Corresponding sides of similar triangles are proportional;" and
6. recognize SAS$\sim$ and SSS$\sim$ as methods of verifying that triangles are similar.

Section 5.3: The Pythagorean Theorem

1. state/prove/apply Theorem 5.5.1 in establishing later theorems;
2. state/apply/prove theorems involving geometric means in the right triangle;
3. state/apply the Pythagorean Theorem and its converse;
4. determine whether (a,b,c) is a Pythagorean Triple; and
5. determine whether a triangle is acute, right, or obtuse based upon the lengths of sides.

Section 5.4: Special Right Triangles

1. state/apply/prove the 45°-45°-90° Theorem;
2. state/apply/prove the 30°-60°-90° Theorem; and

3. recognize/apply the equivalent theorems (Theorems 5.4.3 and 5.4.4).

Section 5.5: Segments Divided Proportionally

1. form an intuitive understanding of the concept "segments divided proportionally";
2. state/apply the definition of segments divided proportionally;
3. state/apply/prove the theorem that establishes that parallel lines determine proportional segments on transversals; and
4. state/apply the theorem, "The angle-bisector in a triangle separates the opposite side into segments whose lengths have the same ratio as the lengths of the sides of the bisected angle."

A Look Beyond Chapter 5: An Unusual Application of Similar Triangles

1. recognize the power of geometry in problem-solving; and
2. become familair with the concept of reflection as applied in this problem.

Chapter Six: Circles

Section 6.1: Circles and Related Segments and Angles

1. become familiar with the terminology (radius, center, chord, arc, etc.) of the circle;
2. state/apply postulates related to the circle;
3. state/apply/prove selected theorems related to the circle; and
4. state/apply methods of measuring central and inscribed angles in the circle.

Section 6.2: More Angle Measures in the Circle

1. state definitions for terms such as tangent and secant (of a circle);
2. recognize when polygons are inscribed in/circumscribed about circles;
3. recognize when circles are inscribed in/circumscribed about polygons; and
4. state/apply/prove theorems that relate angle and arc measures in the circle.

Section 6.3: Line and Segment Relationships in the Circle

1. state/apply/prove theorems relating radii and chords of a circle;
2. recognize/use terminology involving tangent circles;
3. recognize/use terminology involving common tangents to circles; and
4. state/apply/prove theorems involving lengths of chords, tangents, and secants.

Section 6.4: Some Constructions and Inequalities for the Circle

1. state/apply Theorem 6.4.1 (radius drawn to point of tangency is perpendicular to tangent);

2. perform constructions of tangent to circle from point *on* the circle or in *exterior* of circle; and
3. state/apply/prove theorems relating unequal chords, arcs, and central angles of a circle.

Section 6.5: Locus of Points

1. understand/state the definition of the term locus;
2. draw/construct/describe the locus of points for a selected condition(s);
3. recognize the locus of points equidistant from sides of angle (and from endpoints of line segment);
4. recognize/describe the differences between a locus in a plane/space; and
5. verify the locus theorem by establishing two results.

Section 6.6: Concurrence of Lines

1. understand/state the definition of concurrent lines;
2. state/apply/prove the concurrence of the three angle-bisectors of a triangle;
3. state/apply/prove the concurrence of the perpendicular-bisectors of sides of a triangle;
4. state/apply the concurrence of the three altitudes of a triangle; and
5. state/apply the concurrence of the three medians of a triangle.

A Look Beyond Chapter 6: The Value of π

1. know the fact, "The constant ratio of the circumference to the diameter of a circle is π."; and
2. know that some commonly used approximations of π are $\frac{22}{7}$, 3.14, and 3.1416.

Chapter Seven: Areas of Polygons and Circles

Section 7.1: Area and Initial Postulates

1. develop an intuitive understanding of the area concept;
2. distinguish between units of length and units of area measurement;
3. state/apply the initial postulates involving areas of regions; and
4. prove/apply theorems involving area of a square, parallelogram, or triangle.

Section 7.2: Perimeter and Area of Polygons

1. state/apply perimeter formulas for selected polygons;
2. state/apply Heron's Formula for the area of a triangle;
3. state/apply/prove formulas for the areas of trapezoid, rhombus, and kite; and
4. use the ratio between the lengths of corresponding sides of similar polygons to determine the ratio between their areas.

Section 7.3: Regular Polygons and Area

1. determine whether a given polygon can be inscribed in a circle;
2. determine whether a given polygon can be circumscribed about a circle;
3. perform constructions involving inscribed/circumscribed polygons and circles;
4. calculate measure of central angle, radius, and apothem of a regular polygon; and
5. determine the area of a regular polygon by applying the formula $A = \frac{1}{2}aP$.

Section 7.4: The Circumference and Area of a Circle

1. recall that π is the ratio of the circumference to the diameter of a circle;
2. know/apply the formulas $C = \pi d$ and $C = 2\pi r$ for circumference of a circle;
3. memorize the common approximations for π;
4. understand/apply the formula for the length of an arc; and
5. state/apply the formula for the area of a circle.

Section 7.5: More Area Relationships in the Circle

1. understand/apply the formula for the area of a sector;
2. determine the area of a segment of a circle;
3. prove that the area of a triangle with perimeter P and radius r of inscribed circle is given by $A = \frac{1}{2}rP$; and
4. determine the area of a triangle using the formula $A = \frac{1}{2}rP$.

A Look Beyond Chapter 7: Another Look at the Pythagorean Theorem

1. understand/compare the two "area" proofs of the Pythagorean Theorem.

Chapter Eight: Surfaces and Solids

Section 8.1: Prisms, Area and Volume

1. understand intuitively the notion of prism;
2. understand/use terminology (edges, vertices, etc.) related to prisms;
3. determine the lateral area/total area of a prism; and
4. memorize/apply the formula for the volume of a prism.

Section 8.2: Pyramids, Area and Volume

1. understand intuitively the notion of pyramid;
2. understand/use terminology (edges, vertices, etc.) related to pyramids;
3. apply $\ell^2 = a^2 + h^2$, which relates slant height, apothem, and altitude of a regular pyramid;

4. determine the lateral area/total area of a pyramid; and
5. memorize/apply the formula for the volume of a pyramid.

Section 8.3: Cylinders and Cones

1. understand intuitively the notions of cylinder and cone;
2. understand/use terminology related to cylinders and cones;
3. apply $\ell^2 = r^2 + h^2$, which relates slant height, radius, and altitude of a right circular cone;
4. memorize/apply formulas for lateral/total area of a right circular cylinder;
5. memorize/apply formulas for lateral/total area of a right circular cone; and
6. memorize/apply formulas for the volume of a right circular cylinder/cone.

Section 8.4: Polyhedrons and Spheres

1. understand intuitively the notion of polyhedron;
2. know/use the terminology related to a polyhedron/regular polyhedron;
3. verify Euler's Equation $V + F = E + 2$, relating vertices, faces, and edges of a polyhedron;
4. state the five regular polyhedrons;
5. know/apply the term sphere and terminology related to a sphere;
6. memorize/apply formulas for surface area/volume of a sphere; and
7. understand/apply the concept of solid of revolution.

A Look Beyond Chapter 8: Historical Sketch of Descartes

1. recognize Descartes as inventor of the rectangular coordinate system; and
2. recognize/state the geometric figures known as conic sections.

Chapter Nine: Analytic Geometry

Section 9.1: The Rectangular Coordinate System

1. know/use terms related to the rectangular coordinate system;
2. plot/read points in the coordinate system as ordered pairs;
3. find the distance between two points on a vertical/horizontal segment;
4. know/apply/prove the Distance Formula; and
5. know/apply the Midpoint Formula.

Section 9.2 : Graphs of Linear Equations and Slope

1. state/apply the definition of graph of equation;
2. determine/use intercepts in graphing straight lines;

3. know/apply the Slope Formula;
4. determine by sight if a line has positive, negative, 0, or undefined slope; and
5. use slope relationships to determine if lines are parallel/perpendicular.

Section 9.3: Preparing to do Analytic Proofs

1. determine the analytic formula necessary to prove a given statement;
2. prepare the drawing used to complete the analytic proof of a theorem;
3. name/describe general coordinates of vertices for a particular type of geometric figure; and
4. use algebraic relationships to develop geometric relationships in given figures.

Section 9.4: Analytic Proofs

1. develop a logical and orderly plan needed to complete an analytic proof; and
2. construct the analytic proof of a given geometric theorem.

Section 9.5: Equation of Lines

1. use the Slope-Intercept and Point-Slope Forms to find equations of lines;
2. use the equation of a line to draw its graph;
3. use graphs/algebra to solve systems of equations (find points of intersection); and
4. develop analytic proofs for theorems by using equations of lines.

A Look Beyond Chapter 9: The Banach-Tarski Paradox

1. recognize noncollinearity as the reason for the paradox.

Chapter Ten: Introduction to Trigonometry

Section 10.1: The Sine Ratio and Applications

1. define/apply the sine ratio of an acute angle of a right triangle;
2. use a table/calculator to determine the sine ratio of an acute angle;
3. use a table/calculator to determine the acute angle whose sine ratio is known; and
4. understand/apply the notion angle of elevation/depression.

Section 10.2: The Cosine Ratio and Applications

1. define/apply the cosine ratio of an acute angle of a right triangle;
2. use a calculator to determine the cosine ratio of an acute angle;
3. use a calculator to determine the acute angle whose cosine ratio is known; and
4. know/apply/prove the identity $\sin^2\alpha + \cos^2\alpha = 1$.

Section 10.3: The Tangent Ratio and Other Ratios

1. define/apply the tangent ratio of an acute angle of a right triangle;
2. recognize which trigonometric ratio (sine, cosine, tangent) can be used to determine an unknown measure in a right triangle;
3. use a calculator to determine the tangent ratio of an acute angle of a right triangle;
4. use a calculator to determine an acute angle whose tangent ratio is known;
5. state/apply the definitions of the cotangent, secant, and cosecant ratios for an acute angle of a right triangle; and
6. define/determine cot θ, sec θ, and csc θ as reciprocals of tan θ, cos θ, and sin θ respectively.

Section 10.4: More Trigonometric Relationships

1. state/apply the Reciprocal Identities;
2. state/apply the Quotient Relationships;
3. state/apply the Pythagorean Relationships;
4. state/apply the formula $A = \frac{1}{2}bc \sin \alpha$;
5. state/apply the Law of Sines;
6. state/apply the Law of Cosines; and
7. use given measures to decide whether the Law of Sines/Cosines should be used to find an unknown measure in a triangle.

A Look Beyond Chapter 10: Radian Measure of Angles

1. know that a counterclockwise/clockwise rotation corresponds to an angle whose measure is positive/negative;
2. draw/measure angles that have any positive/negative degree measure; and
3. know/apply the fact that $180^\circ = \pi$ radians.

Student Solutions Manual

Contents

Chapter One: Line and Angle Relationships

SECTION 1.1: Statements and Reasoning

1. a. Not a statement
   b. Statement; true
   c. Statement; true
   d. Statement; false

5. Conditional

9. Simple

13. H: The diagonals of a parallelogram are perpendicular.
    C: The parallelogram is a rhombus.

17. First write the statement in "If, then" form.
    If a figure is a square, then it is a rectangle.
    H: A figure is a square.
    C: It is a rectangle.

21. True

25. Induction

29. Intuition

33. Angle 1 looks equal in measure to angle 2.

37. A Prisoner of Society might be nominated for an Academy Award.

41. Angles 1 and 2 are complementary.

45. None

49. Marilyn is a happy person.

SECTION 1.2: Informal Geometry and Measurement

1. $AB < CD$

5. One; none

9. Yes; no; yes

13. Yes; no

17. a. 40° b. 50°

21. Equal

25. Yes

29. $\overline{MN}$ and $\overline{QP}$

33. 22

37. 124

41. $x + 2x + 3 = 72$
    $3x = 69$
    $x = 23$

45. $x + y = 180$
    $x - y = 24$
    $2x = 204$
    $x = 102$; $y = 78$

46. $x + y = 67$
    $x - y = 17$
    $2x = 84$
    $x = 42$; $y = 25$

SECTION 1.3: Early Definitions and Postulates

1. AC

5. $\frac{1}{2}\text{ m} \cdot 3.28\text{ ft/m} = 1.64$ feet

9. $\overleftrightarrow{CD}$ means line CD;
   $\overline{CD}$ means segment CD;
   CD means the measure or length of $\overline{CD}$;
   $\overrightarrow{CD}$ means ray CD with endpoint C.

13. $2x + 1 = 3x - 2$
    $-x = -3$
    $x = 3$; $AM = 7$

17. a. $\overrightarrow{OA}$ and $\overrightarrow{OD}$
    b. $\overrightarrow{OA}$ and $\overrightarrow{OB}$ (Other possible answers.)

21. Planes M and N intersect at $\overleftrightarrow{AB}$.

25. a. C
    b. C
    c. H

29. Given: $\overline{AB}$ as shown

Construct: $\overline{PQ}$ on line *n* so that

PQ = 3(AB)

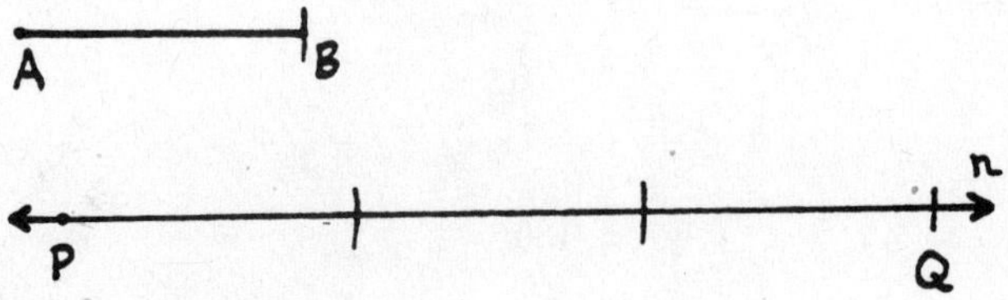

33. Six

SECTION 1.4: Angles and Their Relationships

1. a. Acute

   b. Right

   c. Obtuse

5. Adjacent

9. a. Yes

   b. No

13. m∠FAC + m∠CAD = 180;

    ∠FAC and ∠CAD are supplementary.

17. $2x - 10 + x + 6 = 4(x - 6)$

    $3x - 4 = 4x - 24$

    $20 = x$

    $x = 20$; $m\angle RSV = 4(20 - 6) = 56^\circ$

21. ∠CAB ≅ ∠DAB

25. a. $180 - x$

    b. $180 - (3x - 12)$

    $180 - 3x + 12$; $192 - 3x$

    c. $180 - (2x + 5y)$

    $180 - 2x - 5y$

29. Given: Obtuse ∠MRP

Construct: With $\overrightarrow{OA}$ as one side, an angle ≅ ∠MRP.

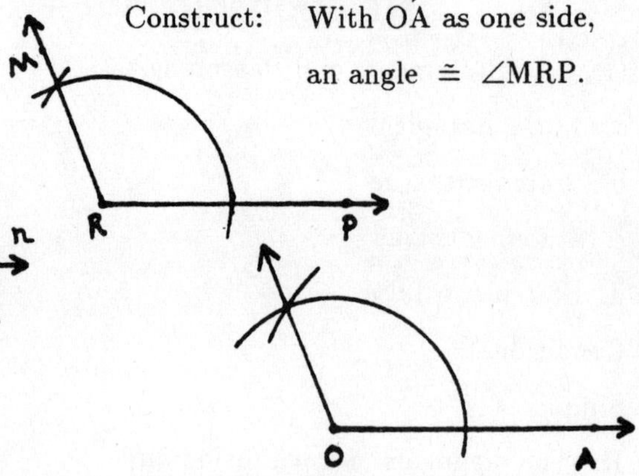

33. For the triangle shown, the angle bisectors have been constructed.

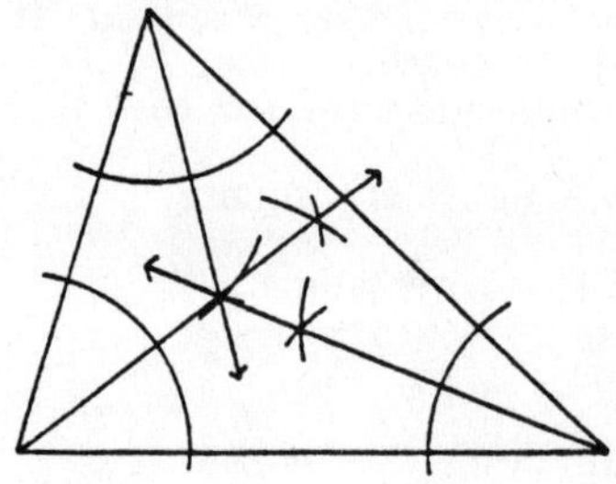

It appears that the angle bisectors meet at one point.

37. a. 90° b. 90° c. Equal

SECTION 1.5: Introduction to Geometric Proof

1. Division Property of Equality or Multiplication Property of Equality

5. Multiplication Property of Equality

9. Angle-Addition Property

13. $\overrightarrow{EG}$ bisects ∠DEF

17. 1. Given
    2. Distributive Property
    3. Addition Property of Equality
    4. Division Property of Equality

21. 1. Given

2. Segment-Addition Postulate

3. Subtraction Property of Equality

25. 1. M-N-P-Q on $\overline{MQ}$; Given

2. Segment-Addition Postulate

3. Segment-Additon Postulate

4. MN + NP + PQ = MQ

29. $(-7)(-2) > 5(-2)$ or $14 > -10$

SECTION 1.6: Relationships: Perpendicular Lines

1. 1. Given

2. If 2 ∠s are ≅ , then they are equal in measure.

3. Angle-Addition Postulate

4. Addition Property of Equality

5. Substitution

6. If 2 ∠s are = in measure, then they are ≅.

5. Given: Point N on line *s*

Construct: Line *m* through N so that $m \perp s$.

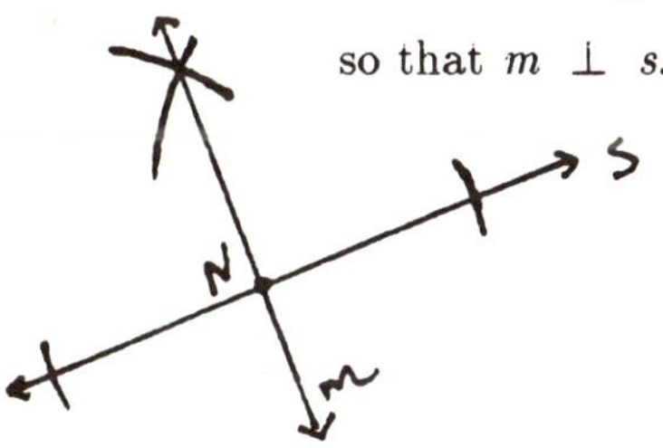

9. Given: Triangle ABC

Construct: The perpendicular bisectors of each side, $\overline{AB}$, $\overline{AC}$, and $\overline{BC}$.

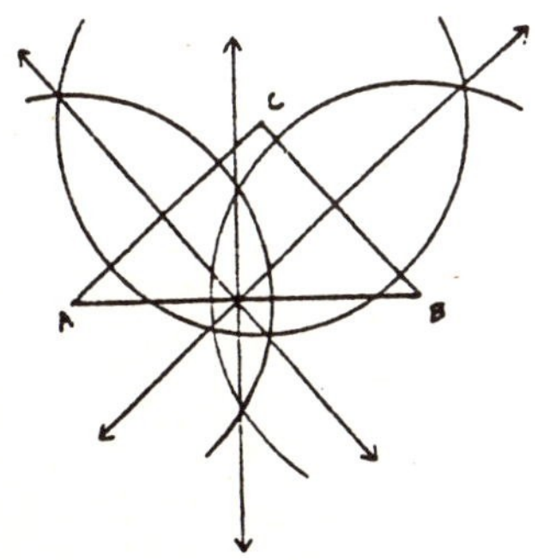

13. No; Yes; No

17. No; Yes; Yes

21.

1. M-N-P-Q on $\overline{MQ}$	1. Given
2. MN + NQ = MQ	2. Segment-Addition Postulate
3. NP + PQ = NQ	3. Segment-Addition Postulate
4. MN + NP + PQ = MQ	4. Substitution

SECTION 1.7: The Formal Proof of a Theorem

1. H: A line segment is bisected.

C: Each of the equal segments has half the length of the original segment.

5. H: Each is a right angle.

C: Two angles are congruent.

9. Given: $\overleftrightarrow{AB} \perp \overleftrightarrow{CD}$

Prove: ∠AEC is a right angle

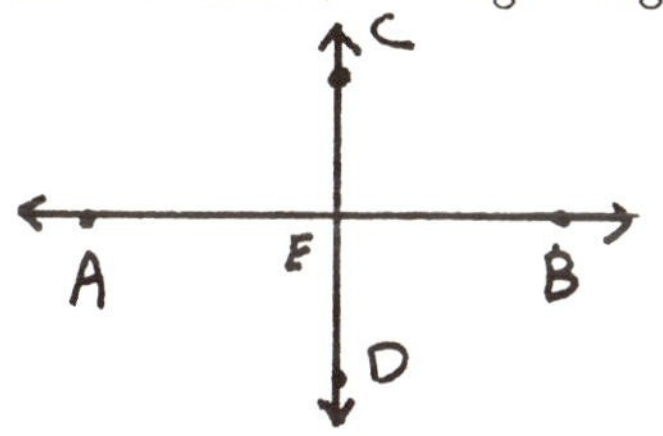

Figure for Exercise 9

13. Given: Lines ℓ and *m*

Prove: ∠1 ≅ ∠2 and ∠3 ≅ ∠4

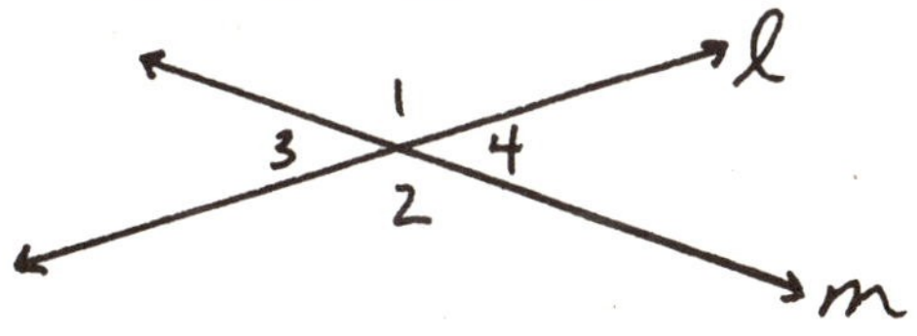

17. m∠1 = m∠3

3x + 10 = 4x − 30

x = 40; m∠1 = 130°

21. 1. Given

2. If 2 $\angle$s are comp., then the sum of their measures is 90.

3. Substitution

4. Subtraction Property of Equality

5. If 2 $\angle$s are = in measure, then they are $\cong$.

25. 1. Given

2. $\angle$ABC is a right $\angle$

3. The measure of a rt. $\angle$ = 90.

4. Angle-Addition Postulate

6. $\angle 1$ is comp. to $\angle 2$

29. The supplement of an acute angle is obtuse.

Given: $\angle 1$ is supp to $\angle 2$

$\angle 2$ is an acute $\angle$

Prove: $\angle 1$ is an obtuse $\angle$

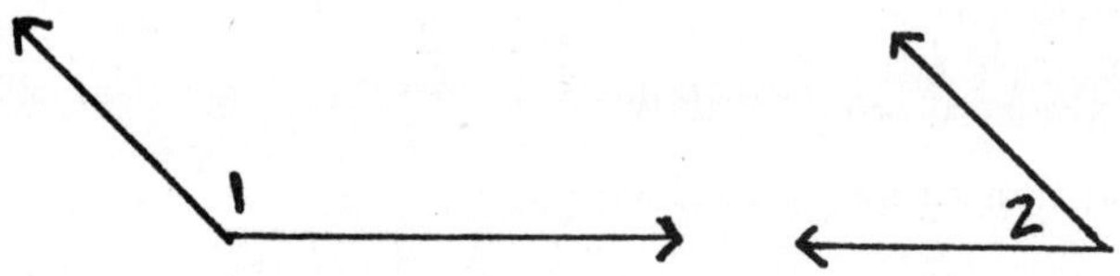

STATEMENTS	REASONS
1. $\angle 1$ is supp to $\angle 2$	1. Given
2. $m\angle 1 + m\angle 2 = 180$	2. If 2 $\angle$s are supp., the sum of their measures is 180.
3. $\angle 2$ is an acute $\angle$	3. Given
4. $m\angle 2 = x$ where $0 < x < 90$	4. The measure of an acute $\angle$ is between 0 and 90.
5. $m\angle 1 + x = 180$	5. Substitution (#4 into #2)
6. x is positive $\therefore m\angle 1 < \angle 180$	6. If $a + p_1 = b$ and p_1 is positive, then $a < b$.
7. $m\angle 1 = 180 - x$	7. Substitution Prop. of Eq. (#5)
8. $-x < 0 < 90 - x$	8. Subtraction Prop. of Ineq. (#4)
9. $90 - x < 90 < 180 - x$	9. Addition Prop. of Ineq. (#8)
10. $90 - x < 90 < m\angle 1$	10. Substitution (#7 into #9)
11. $90 < m\angle 1 < 180$	11. Transitive Prop. of Ineq. (#6 & #10)
12. $\angle 1$ is an obtuse $\angle$	12. If the measure of an angle is between 90 and 180, then the $\angle$ is obtuse.

CHAPTER 1: Review

1. Undefined terms, defined terms, axioms or postulates, theorems.
2. Induction, deduction, intuition
3. 1. Names the term being defined
   2. Places the term into a set or category
   3. Distinguishes the term from other terms in the same category
   4. Reversible
4. Intuition
5. Induction
6. Deduction
7. H: The diagonals of a trapezoid are equal in length.
   C: The trapezoid is isosceles.
8. H: The parallelogram is a rectangle.
   C: The diagonals of a parallelogram are congruent.
9. No conclusion
10. Jody smithers has a college degree.
11. Angle A is a right angle.
12. C
13. ∠RST, ∠S, more than 90°
14. Diagonals are ⊥ and they bisect each other.
15.

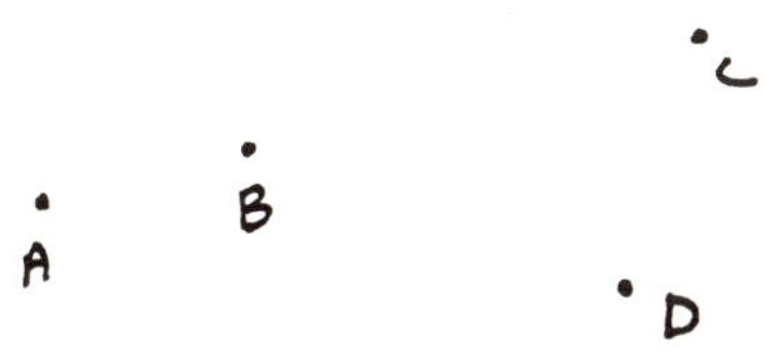

16.

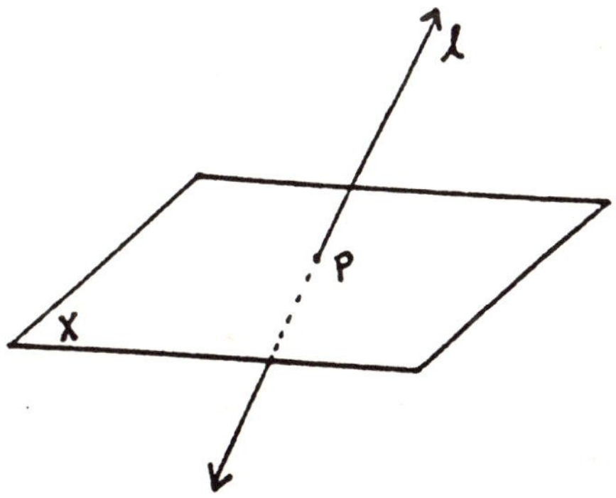

17.

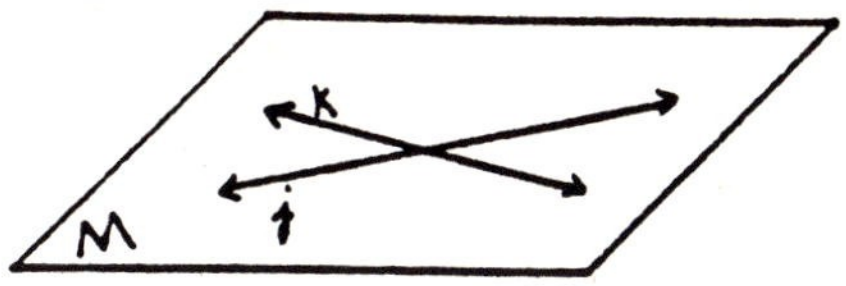

18. $2x + 15 = 3x - 2$

    $17 = x$

    $x = 17$; $m\angle ABC = 98°$

19. $2x + 5 + 3x - 4 = 86$

    $5x + 1 = 86$

    $5x = 85$

    $x = 17$; $m\angle DBC = 47$

20. $3x - 1 = 4x - 5$

    $4 = x$

    $x = 4$; $AB = 22$

21. $4x - 4 + 5x + 2 = 25$

    $9x - 2 = 25$

    $9x = 27$

    $x = 3$; $MB = 17$

22. $2 \cdot CD = BC$

    $2(2x + 5) = x + 28$

    $4x + 10 = x + 28$

    $3x = 18$

    $x = 6$; $AC = BC = 6 + 28 = 34$

23. $7x - 21 = 3x + 7$

    $4x = 28$

    $x = 7$

    $m\angle 3 = 49 - 21 = 28$

    $\therefore m\angle FMH = 180 - 28 = 152$

24. $4x + 1 + x + 4 = 180$

    $5x + 5 = 180$

    $5x = 175$

    $x = 35$

    $m\angle 4 = 35 + 4 = 39$

25. $2x - 6 + 3(2x - 6) = 90$

$2x - 6 + 6x - 18 = 90$

$8x - 24 = 90$

$8x = 114$

$x = 14\frac{1}{4}$

$m\angle EFH = 3(2x - 6) = 3(28\frac{1}{2} - 6)$

$= 3 \cdot 22\frac{1}{2}$

$= 67\frac{1}{2}$

26. $x + (40 + 4x) = 180$

$5x + 40 = 180$

$5x = 140$

$x = 28$

The angles measure 28 and 152.

27. a. $2x + 3 + 3x - 2 + x + 7 = 6x + 8$

b. $6x + 8 = 32$

$6x = 24$

$x = 4$

c. $2x + 3 = 2(4) + 3 = 11;$

$3x - 2 = 3(4) - 2 = 10;$

$x + 7 = 4 + 7 = 11$

28. The measure of angle 3 is less than 50.

29. The 4 foot board is 48 inches. Subtract 6 inches on each end leaving 36 inches.

$4(n - 1) = 36$

$4n - 4 = 36$

$4n = 40$

$n = 10$

$\therefore$ 10 pegs will fit on the board.

30. S

31. S

32. A

33. S

34. N

35. 2. $\angle 4 \cong \angle P$

3. $\angle 1 \cong \angle 4$

4. If 2 $\angle$s are $\cong$, then their measures are =.

5. Given

6. $m\angle 2 = m\angle 3$

7. $m\angle 1 + m\angle 2 = m\angle 4 + m\angle 3$

8. Angle-Addition Postulate

9. Substitution

10. $\angle TVP \cong \angle MVP$

36. Given: $\overline{KF} \perp \overline{FH}$

∠JHK is a right ∠

Prove: ∠KFH ≐ ∠JHF

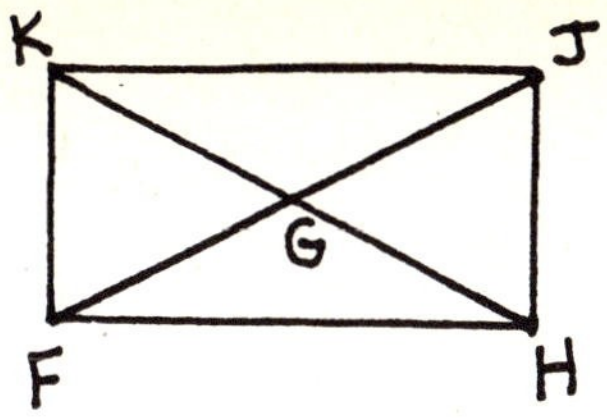

STATEMENTS	REASONS
1. $\overline{KF} \perp \overline{FH}$	1. Given
2. ∠KFH is a rt. ∠	2. If 2 segments are ⊥, then they form a rt ∠.
3. ∠JHF is a rt ∠	3. Given
4. ∠KFH ≐ ∠JHF	4. Any two right ∠s are ≐.

37. Given: $\overline{KH} \doteq \overline{FJ}$

G is the midpoint of both $\overline{KH}$ and $\overline{FJ}$

Prove: $\overline{KJ} \doteq \overline{GJ}$

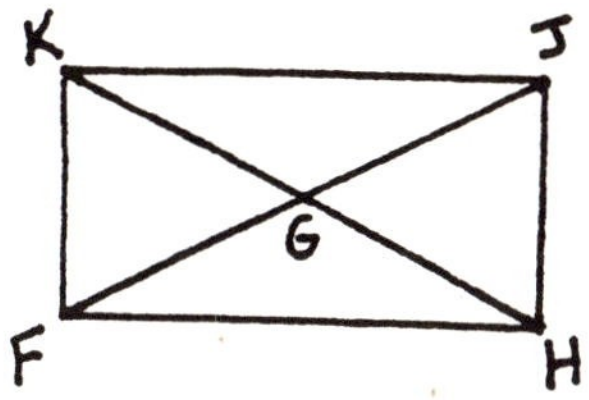

STATEMENTS	REASONS
1. $\overline{KH} \doteq \overline{FJ}$ G is the midpoint of both $\overline{KH}$ and $\overline{FJ}$	1. Given
2. $\overline{KG} \doteq \overline{GJ}$	2. If 2 segments are ≐, then their midpoints separate these segments into 4 ≐ segments.

38. Given: $\overline{KF} \perp \overline{FH}$

Prove: ∠KFJ is comp. to ∠ JFH

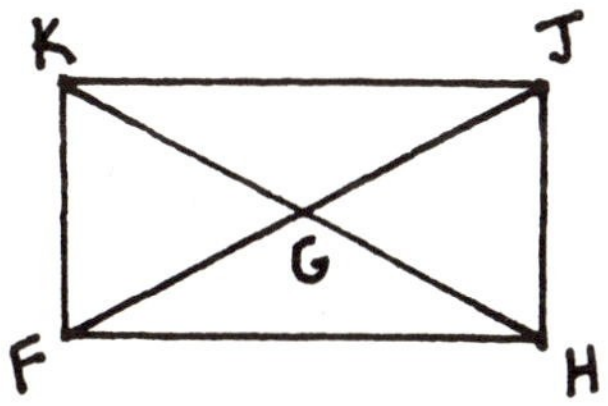

STATEMENTS	REASONS
1. $\overline{KF} \perp \overline{FH}$	1. Given
2. ∠KFJ is comp. to ∠JFH	2. If the exterior sides of 2 adjacent ∠s form ⊥ rays, then these ∠s are comp.

39. Given: $\angle 1$ is comp. to $\angle M$

$\angle 2$ is comp. to $\angle M$

Prove: $\angle 1 \doteq \angle 2$

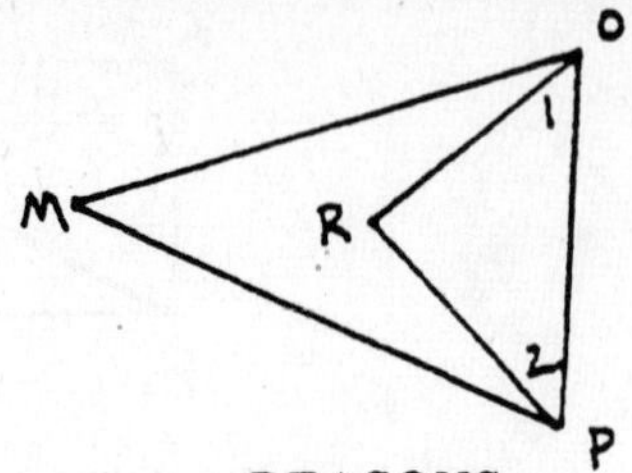

STATEMENTS	REASONS
1. $\angle 1$ is comp. to $\angle M$	1. Given
2. $\angle 2$ is comp. to $\angle M$	2. Given
3. $\angle 1 \doteq \angle 2$	3. If 2 $\angle$s are comp. to the same $\angle$, then these angles are $\doteq$.

40. Given: $\angle MOP \doteq \angle MPO$

$\overrightarrow{OR}$ bisects $\angle MOP$

$\overrightarrow{PR}$ bisects $\angle MPO$

Prove: $\angle 1 \doteq \angle 2$

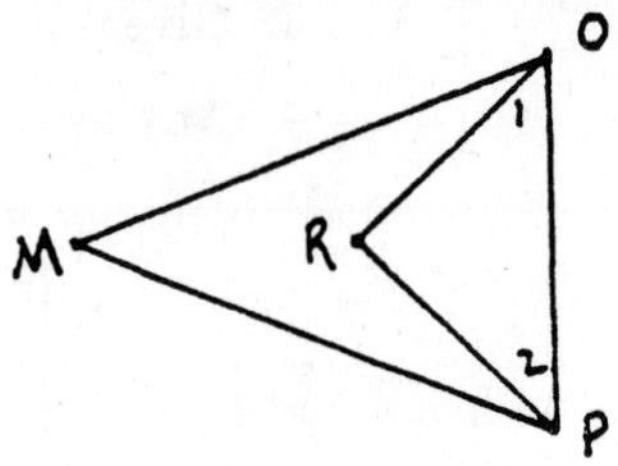

STATEMENTS	REASONS
1. $\angle MOP \doteq \angle MPO$	1. Given
2. $\overrightarrow{OR}$ bisects $\angle MOP$ $\overrightarrow{PR}$ bisects $\angle MPO$	2. Given
3. $\angle 1 \doteq \angle 2$	3. If 2 $\angle$s are $\doteq$, then their bisectors separate these $\angle$s into four $\doteq$ $\angle$s.

41. Given: $\angle 4 \doteq \angle 6$

Prove: $\angle 5 \doteq \angle 6$

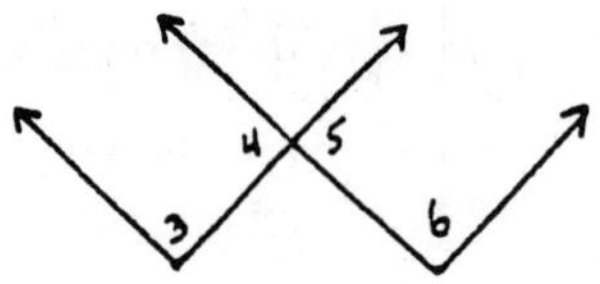

STATEMENTS	REASONS
1. $\angle 4 \doteq \angle 6$	1. Given
2. $\angle 4 \doteq \angle 5$	2. If 2 angles are vertical $\angle$s, then they are $\doteq$.
3. $\angle 5 \doteq \angle 6$	3. Transitive Prop.

42. Given: Figure as shown

Prove: $\angle 4$ is supp. to $\angle 2$

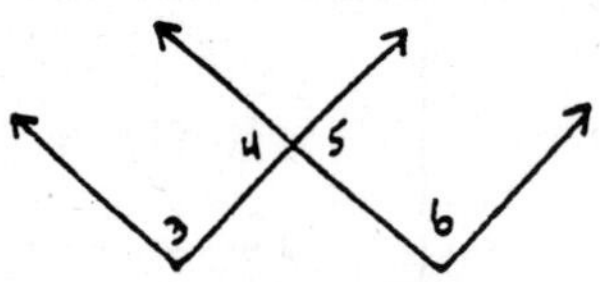

STATEMENTS	REASONS
1. Figure as shown	1. Given
2. $\angle 4$ is supp. to $\angle 2$	2. If the exterior sides of 2 adjacent $\angle$s form a line, then the $\angle$s are supp.

43. Given: $\angle 3$ is supp. to $\angle 5$

$\angle 4$ is supp. to $\angle 6$

Prove: $\angle 3 \cong \angle 6$

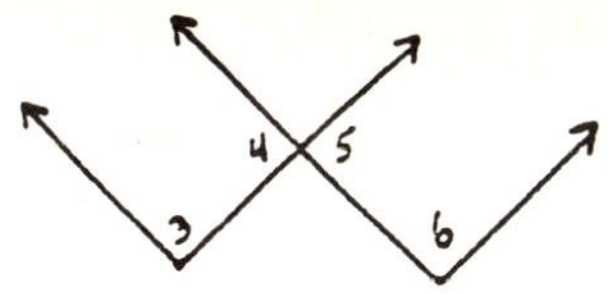

STATEMENTS	REASONS
1. $\angle 3$ is supp to $\angle 5$ $\angle 4$ is supp to $\angle 6$	1. Given
2. $\angle 4 \cong \angle 5$	2. If 2 lines intersect, the vertical angles formed are $\cong$.
3. $\angle 3 \cong \angle 6$	3. If 2 $\angle$s are supp to congruent angles, then these angles are $\cong$.

44. Given: $\overline{VP}$

Construct: $\overline{VW}$ such that $VW = 4 \cdot VP$

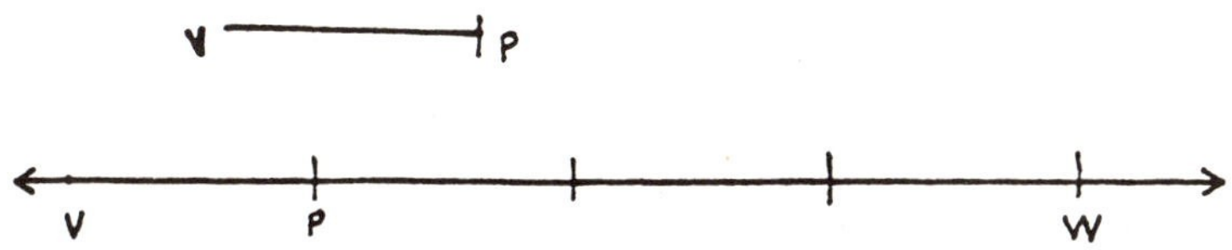

45. Construct a 135° angle.

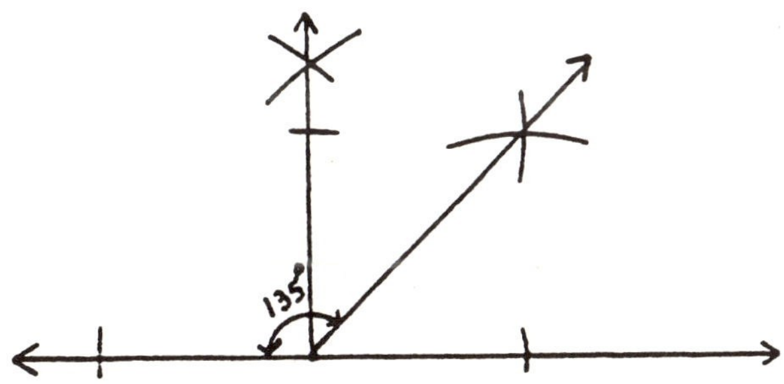

46. Given: Triangle PQR

Construct: The three angle bisectors.

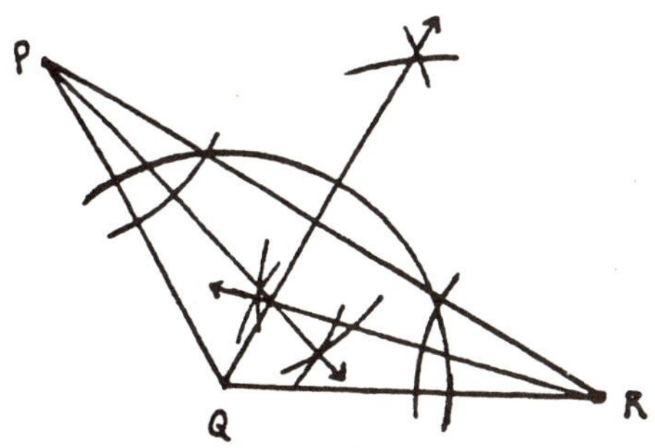

It appears that the three angle bisectors meet at one point inside the triangle.

47. Given: $\overline{AB}$, $\overline{BC}$, and $\angle B$ as shown

Construct: Triangle ABC

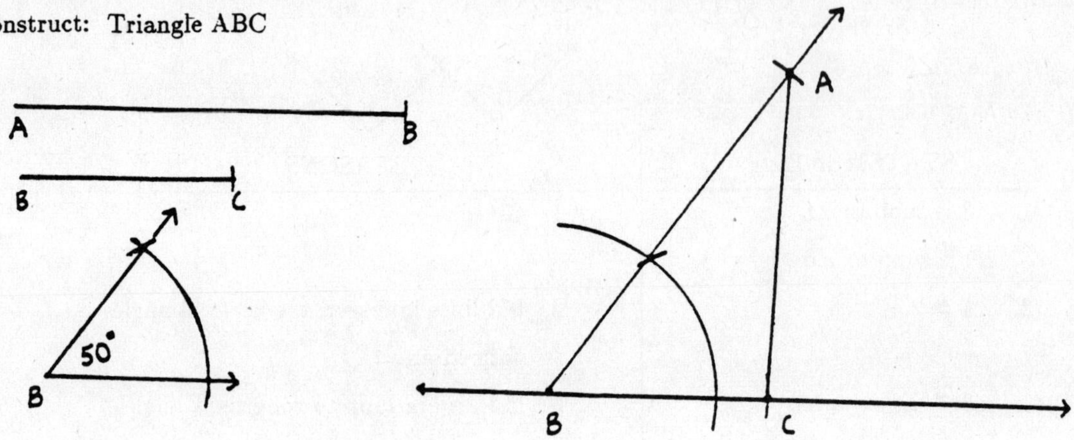

48. Given: $m\angle B = 50°$

Construct: An angle whose measure is 20°.

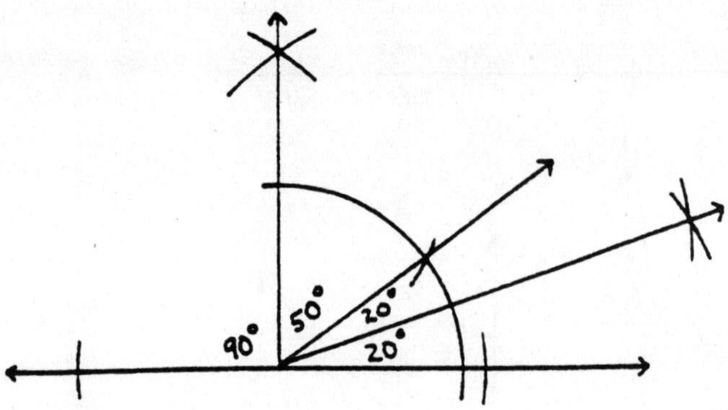

Chapter Two: Parallel Lines

SECTION 2.1: The Parallel Postulate and Special Angles

1. a. $108°$ b. $72°$

5. a. No b. Yes c. Yes

9. a. $m\angle 3 = 87$; $\angle 3$ is vertical to $\angle 2$
   b. $m\angle 6 = 87$;
   $\angle 6$ corresponds to $\angle 2$
   c. $m\angle 1 = 93$;
   $\angle 1$ is supplementary to $\angle 2$
   d. $m\angle 7 = 87$;
   $\angle 7$ corresponds to $\angle 3$.

13. a. $m\angle 2 = 68$; $\angle 2$ is supp. to $\angle 1$.
   b. $m\angle 4 = 112$; $\angle 4$ is vertical to $\angle 1$
   c. $m\angle 5 = 112$; $\angle 5$ is an alternate interior $\angle$ to $\angle 4$.
   d. $m\angle MOQ = 34$;
   $m\angle MON = m\angle 2 = 68$;
   $m\angle MOQ = \frac{1}{2}$of $m\angle MON = 34$.

17. Angles 3 and 5 are supp. because they are interior angles on the same side of the transversal. Angles 5 and 6 are also supp. This leads to a system of 2 equations with 2 variables.
   $(6x + y) + (8x + 2y) = 180$
   $(8x + 2y) + (4x + 7y) = 180$
   Simplifying yields
   $14x + 3y = 180$
   $12x + 9y = 180$

   Dividing the 2nd equation by -3 gives
   $14x + 3y = 180$
   $-4x - 3y = -60$

   Addition gives
   $10x = 120$
   $x = 12$
   Using $14x + 3y = 180$ and $x = 12$ we get
   $14(12) + 3y = 180$
   $168 + 3y = 180$
   $3y = 12$
   $y = 4$
   $m\angle 6 = 4(12) + 7(4) = 76$
   $\therefore$ $m\angle 7$ also $= 76$.

21. Given: $\overleftrightarrow{CE} \parallel \overleftrightarrow{DF}$; trans. $\overleftrightarrow{AB}$;
$\overrightarrow{CX}$ bisects $\angle ACE$ and
$\overrightarrow{DE}$ bisects $\angle CDF$.
Prove: $\angle 1 \doteq \angle 3$

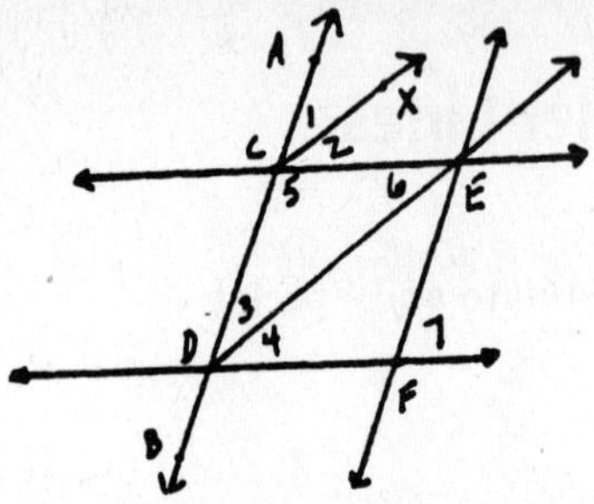

STATEMENTS	REASONS
1. $\overleftrightarrow{CE} \parallel \overleftrightarrow{DF}$; trans. $\overleftrightarrow{AB}$	1. Given
2. $\angle ACE \doteq \angle ADF$	2. If 2 $\parallel$ lines are cut by a trans., then the corresponding $\angle$s are $\doteq$.
3. $\overrightarrow{CX}$ bisects $\angle ACE$ $\overrightarrow{DE}$ bisects $\angle CDF$	3. Given
4. $\angle 1 \doteq \angle 3$	4. If 2 $\angle$s are $\doteq$, then their bisectors separate these $\angle$s into four $\doteq$ $\angle$s.

25. a. $\angle 4 \doteq \angle 2$ and $\angle 5 \doteq \angle 3$
b. 180
c. 180

29. No

33. Given: Triangle MNQ with obtuse $\angle MNQ$
Construct: $\overline{MR} \perp \overline{NQ}$
(Hint: Extend $\overline{NQ}$)

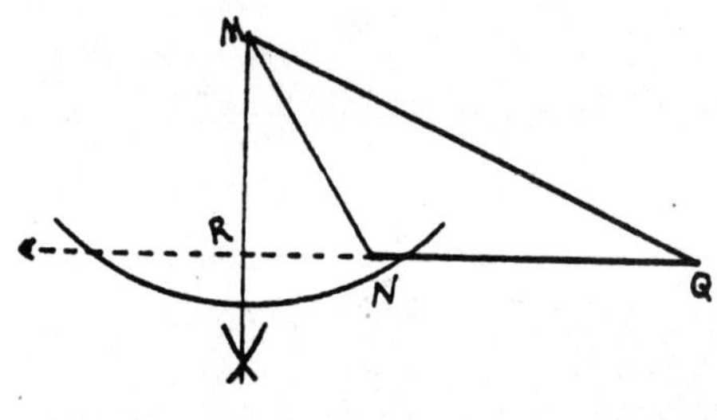

1 . If Juan wins the state lottery, then he will be rich.

Converse: If Juan is rich, then he won the state lottery. FALSE.

Inverse: If Juan does not win the state lottery, then he will not be rich. FALSE.

Contrapositive: If Juan is not rich, then he did not win the state lottery. TRUE.

5. No conclusion.

9 . (a) (b) (e)

13. Given: $\angle AOD \not\doteq \angle AFE$

Prove: $\overleftrightarrow{DC} \nparallel \overleftrightarrow{EG}$

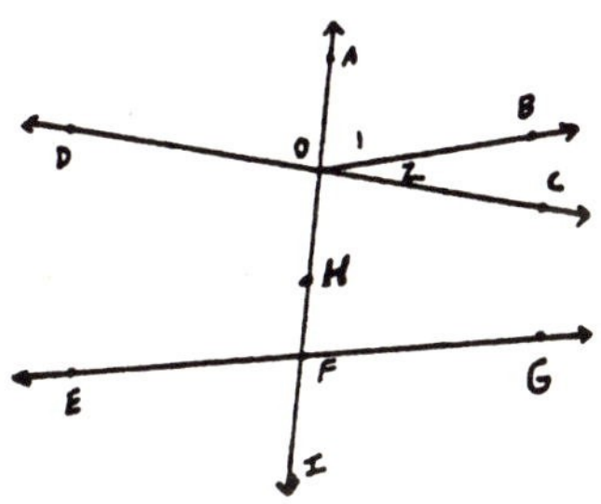

Proof:

Assume that $\overleftrightarrow{DC} \parallel \overleftrightarrow{EG}$. If they are $\parallel$, then $\angle AOD \doteq \angle AFE$ because they are corresponding angles. But this contradicts the Given information. Therefore, our assumption is false and $\overleftrightarrow{DC} \nparallel \overleftrightarrow{EG}$

17. Assume that the angles are vertical angles. If they are vertical angles, then they are congruent. But this contradicts the hypothesis that the two angles are not congruent. Hence, our assumption must be false and the angles are not vertical angles.

21. Given: M is a midpoint of $\overline{AB}$

Prove: M is the only midpoint of $\overline{AB}$.

Proof: If M is a midpoint of $\overline{AB}$, then $AM = \frac{1}{2} \cdot AB$. Assume that N is also a midpoint of $\overline{AB}$ so that $AN = \frac{1}{2} \cdot AB$. By substitution $AM = AN$.

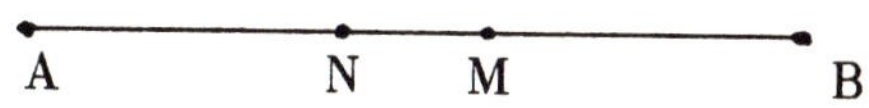

By the Segment-Addition Postulate, $AM = AN + NM$. Using substitution again, $AN + NM = AN$. Subtracting gives $NM = 0$. But this contradicts the Ruler Postulate which states that the measure of a line segment is a positive number. Therefore, our asumption is wrong and M is the only midpoint for $\overline{AB}$.

SECTION 2.3: Proving Lines Parallel

1. $\ell \parallel m$

5. $\ell \nparallel m$ (Sum = 181°)

9. None

13. None

17. 1. Given

 2. If 2 $\angle$s are comp. to the same $\angle$, then they are $\doteq$.

 3. $\overline{BC} \parallel \overline{DE}$

21. Given: $\overrightarrow{DE}$ bisects $\angle CDA$;

$\angle 3 \doteq \angle 1$

Prove: $\overline{ED} \parallel \overline{AB}$

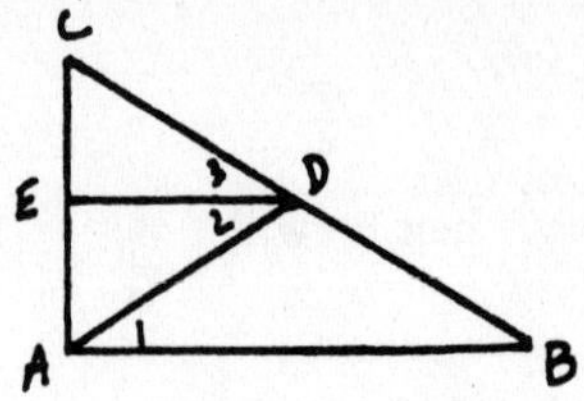

STATEMENTS	REASONS
1. $\overrightarrow{DE}$ bisects $\angle CDA$	1. Given
2. $\angle 2 \doteq \angle 3$	2. If a ray bisects an $\angle$, then it forms 2 $\doteq$ $\angle$s.
3. $\angle 3 \doteq \angle 1$	3. Given
4. $\angle 2 \doteq \angle 1$	4. Transitive for $\doteq$.
5. $\overline{ED} \parallel \overline{AB}$	5. If 2 lines are cut by a trans. so that alternate interior $\angle$s are $\doteq$, then these lines are $\parallel$.

25. $(x + 1)(x + 4) + 16(x + 3) - (x^2 - 2) = 180$

$x^2 + 5x + 4 + 16x + 48 - x^2 + 2 = 180$

$21x + 54 = 180$

$21x = 126$

$x = 6$

29. If two lines are parallel to the same line, then these lines are parallel to each other.

Given: $a \parallel b$ and $c \parallel b$

with trans. t

Prove: $a \parallel c$

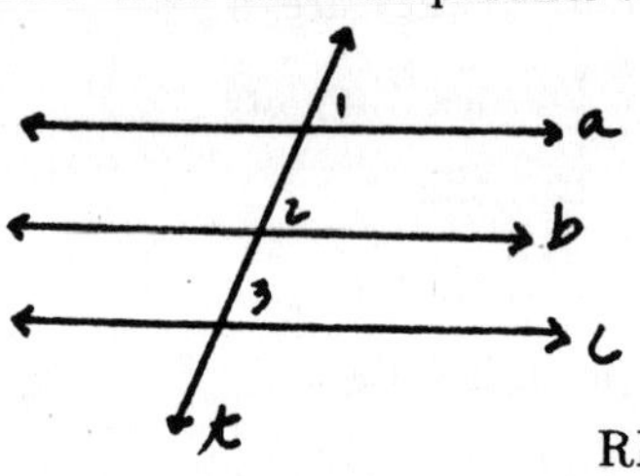

STATEMENTS	REASONS
1. $a \parallel b$ with trans. t	1. Given
2. $\angle 1 \doteq \angle 2$	2. If 2 $\parallel$ lines are cut by a trans., then the corresponding $\angle$s are $\doteq$.
3. $c \parallel b$ with trans. t	3. Given
4. $\angle 2 \doteq \angle 3$	4. Same as (2)
5. $\angle 1 \doteq \angle 3$	5. Transitive for $\doteq$.
6. $a \parallel c$	6. If 2 lines are cut by a trans. so that corr. $\angle$s are $\doteq$, then these lines are $\parallel$.

33. Given: $\overline{BC}$ and A not on $\overline{BC}$

Construct: The segment from A $\perp$ $\overline{BC}$

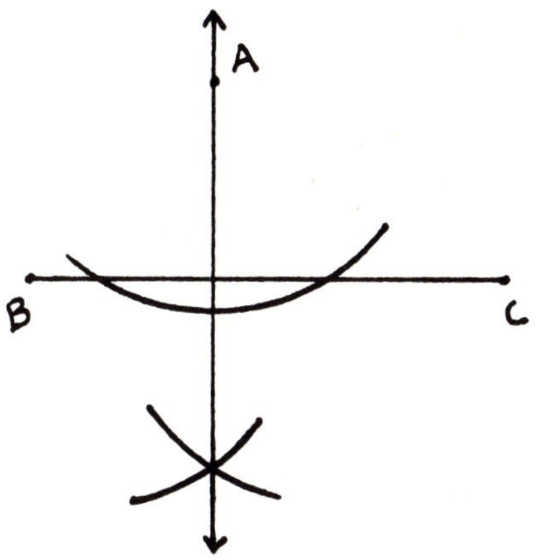

SECTION 2.4: The Angles of a Triangle

1. $m\angle C = 75^\circ$

5. a. Equilateral $\triangle$ b. Isosceles $\triangle$

9. If 2 $\angle$s of one $\triangle$ are $\cong$ to 2 $\angle$s of another $\triangle$, then the third $\angle$s are $\cong$.

13. $m\angle 2 = 57.7$; $m\angle 3 = 80.8$; $m\angle 4 = 41.5$

17. 40

21. $3x = x + 90$ $(m\angle 3 = m\angle 1 + m\angle C)$

$2x = 90$

$x = 45$

$x + y = 90$; $y = 45$

25. Given: $\overline{MN}$ and $\overline{PQ}$ intersect at K as shown;

$\angle M \doteq \angle Q$

Prove: $\angle P \doteq \angle N$

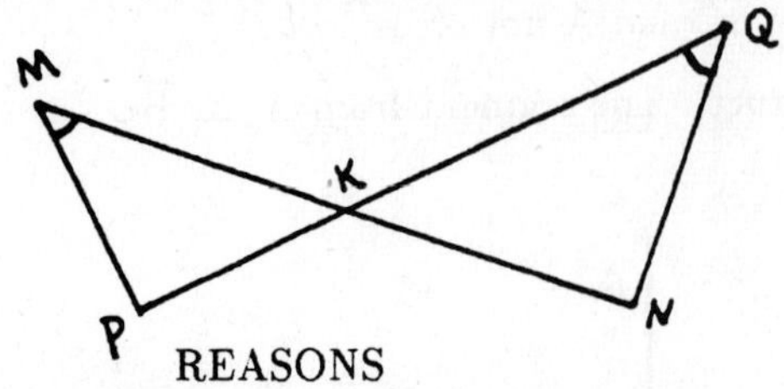

STATEMENTS	REASONS
1. $\overline{MN}$ and $\overline{PQ}$ intersect at K; $\angle M \doteq \angle Q$	1. Given
2. $\angle MKP \doteq \angle QKN$	2. If 2 lines intersect, the vertical $\angle$s formed are $\doteq$.
3. $\angle P \doteq \angle N$	3. If 2 $\angle$s of one $\triangle$ are $\doteq$ to 2 $\angle$s of another $\triangle$, then the third $\angle$s are also $\doteq$.

29. 44

33. 75

37. The measure of an exterior angle of a triangle equals the sum of measures of the two nonadjacent interior angles.

Given: $\triangle ABC$ with ext. $\angle BCD$

Prove: $m\angle BCD = m\angle A + m\angle B$

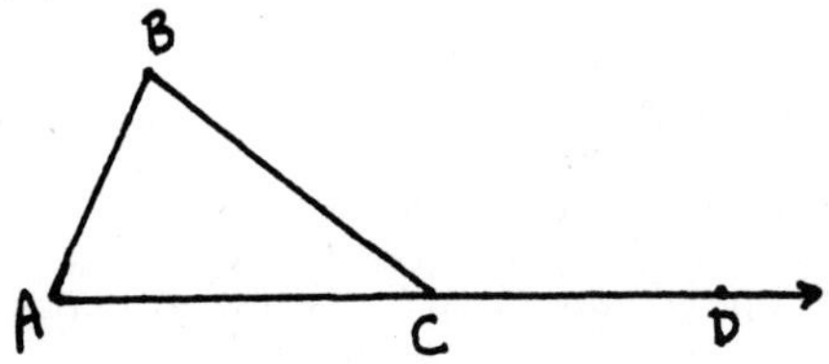

STATEMENTS	REASONS
1. $\triangle ABC$ with ext. $\angle BCD$	1. Given
2. $m\angle A + m\angle B + m\angle BCA = 180$	2. The sum of the measures of the $\angle$s in a $\triangle$ is 180.
3. $\angle BCA$ is supp. to $\angle BCD$	3. If the exterior sides of two adjacent angles form a straight line, then these angles are supp.
4. $m\angle BCA + m\angle BCD = 180$	4. If 2 $\angle$s are supp., then the sum of their measures is 180.
5. $m\angle BCA + m\angle BCD = m\angle A + m\angle B + m\angle BCA$	5. Substitution
6. $m\angle BCD = m\angle A + m\angle B$	6. Subtraction Prop. of Equality

41. $2b = m\angle M + 2a$

$(m\angle RPM = m\angle M + m\angle MNP)$

$\therefore\ m\angle M = 2b - 2a$

$b = 42 + a$

$(m\angle QPR = m\angle Q + m\angle QNP)$

$m\angle M = 2(42 + a) - 2a$ (Substitution)

$m\angle M = 84 + 2a - 2a$

$m\angle M = 84$

SECTION 2.5: Convex Polygons

1. Increase

For #5 use $D = \frac{n(n - 3)}{2}$

5. a. 5 b. 35

For #9 use $I = \frac{180(n - 2)}{n}$

9. a. 90 b. 150

For #13 use $S = 180(n - 2)$

13. a. 7 b. 9

For #17 use $n = \frac{360}{E}$

17. a. 15 b. 20

21. Given: Quad. RSTQ with ext. ∠s at R and T

Prove: m∠1 + m∠2 = m∠3 + m∠4

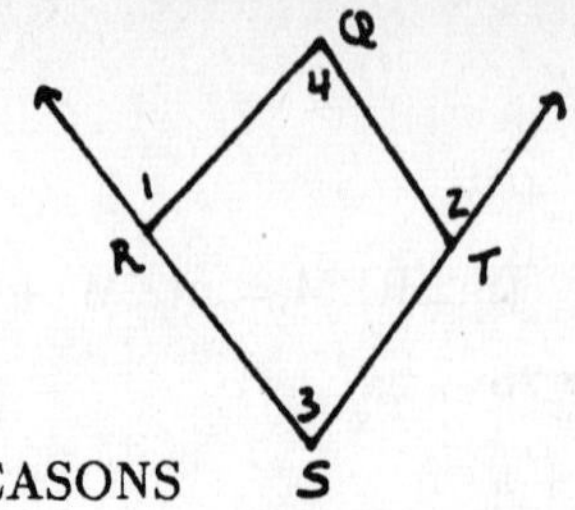

STATEMENTS	REASONS
1. Quad. RSTQ with ext. ∠s at R and T	1. Given
2. m∠QRS + m∠3 + m∠STQ + m∠4 = 360	2. The sum of the measures of the angles in a quad. is 360.
3. ∠1 is supp. to ∠QRS and ∠2 is supp. to ∠QTS	3. If the exterior sides of two adjacent angles form a straight line, then these angles are supp.
4. m∠1 + m∠QRS = 180 and m∠2 + m∠QTS = 180	4. If 2 ∠s are supp., then the sum of their measures is 180.
5. m∠1 + m∠QRS + m∠2 + m∠QTS = 360	5. Addition Property of Equality
6. m∠QRS + m∠3 + m∠STQ + m∠4 = m∠1 + m∠QTS + m∠2 + m∠QTS	6. Substitution
7. m∠1 + m∠2 = m∠3 + m∠4	7. Subtraction Property of Equality

25. (a) 90, 90, 120, 120, 120

(b) 90, 90, 90, 135, 135

29. The resulting polygon is also a REGULAR polygon.

33. a. n − 3

b. $\frac{n(n - 3)}{2}$

1. (a) $\overline{BC} \parallel \overline{AD}$

   (b) $\overline{AB} \parallel \overline{CD}$

2. $m\angle 3 = 110$

3. $2x + 17 = 5x - 94$

   $111 = 3x$

   $37 = x$

4. $m\angle A = 50$ (corresponds to $\angle DCE$)

   $\therefore\ m\angle BCA = 55$

   $m\angle BCD = 75$ and $m\angle D = 75$.

   $m\angle DEF = 50 + 75 = 125$

5. $130 + 2x + y = 180$

   $150 + 2x - y = 180$

   $2x + y = 50$

   $\underline{2x - y = 30}$

   $4x \quad = 80$

   $x = 20$

   $130 + 2(20) + y = 180$

   $130 + 40 + y = 180$

   $y = 10$

6. $2x + 15 = x + 45$

   $1x = 30$

   $3y + 30 + 45 = 180$

   $3y + 75 = 180$

   $3y = 105$

   $y = 35$

7. $\overline{AE} \parallel \overline{BF}$

8. None

9. $\overline{BE} \parallel \overline{CF}$

10. $\overline{BE} \parallel \overline{CF}$

11. $\overline{AC} \parallel \overline{DF}$ and $\overline{AE} \parallel \overline{BF}$

12. $x = 120$ (corr.$\angle$);

    $x = y + 50$

    $120 = y + 50$

    $y = 70$

13. $x = 32;\ y = 30$

14. $2x - y = 3x + 2y$

    $-1x - 3y = 0$ (Multiply by 3)

    $3x - y = 80$

    $-3x - 9y = 0$

    $\underline{3x - y = 80}$

    $-10y = 80$

    $y = -8$

    $3x + 8 = 80$

    $3x = 72$

    $x = 24$

15. $2a + 2b + 100 = 180$

    $2a + 2b = 80$

    $a + b = 40$ ($\div$ by 2)

    $(a + b) + x = 180$

    $40 + x = 180$

    $x = 140$

16. $x^2 - 12 = x(x - 2)$

    $x^2 - 12 = x^2 - 2x$

    $-12 = -2x$

    $x = 6$

17. $x^2 - 3x + 4 + 17x - x^2 - 5 = 111$

$$14x - 1 = 111$$
$$14x = 112$$
$$x = 8$$

$m\angle 3 = 69$; $m\angle 4 = 67$; $m\angle 5 = 44$

18. $3x + y + 5x + 10 = 180$

$3x + y = 5y + 20$

$8x + y = 170$ (Multiply by 4)

$3x - 4y = 20$

$32x + 4y = 680$

$\underline{3x - y = 80}$

$35x \quad = 700$

$x = 20$

$8(20) + y = 170$

$160 + y = 170$

$y = 10$

$m\angle C = 5(10) + 20 = 70 \therefore m\angle B = 110.$

19. S

20. N

21. N

22. S

23. S

24. A

25.

Number of sides	8	12	20	15	10	16	180
Measure of each ext. ∠	45	30	18	24	36	22.5	2
Measure of each int. ∠	135	150	162	156	144	157.5	178
Number of diagonals	20	54	170	90	35	104	15930

26.

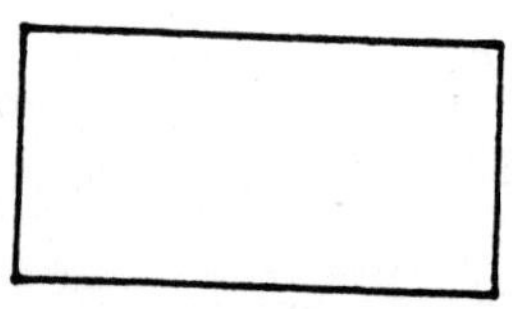

27.

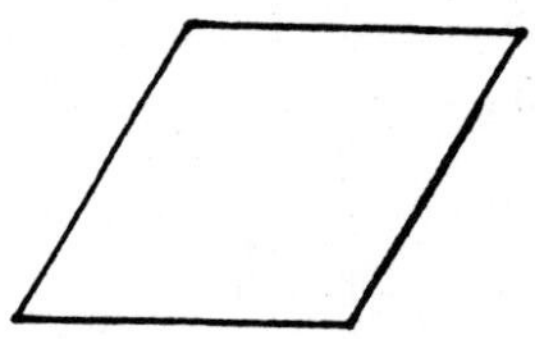

28. Not possible

29.

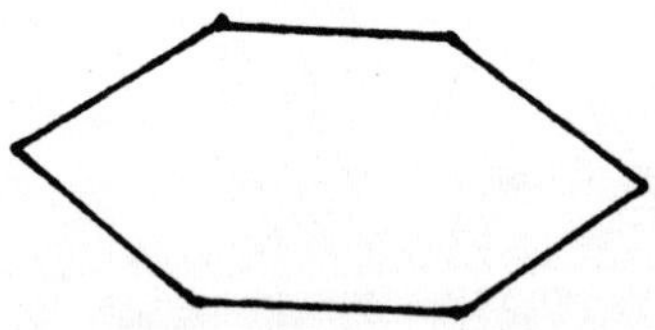

30. STATEMENT: If 2 angles are right angles, then the angles are congruent.

CONVERSE: If 2 angles are congruent, then the angles are right angles.

INVERSE: If 2 angles are not right angles, then the angles are not congruent.

CONTRAPOSITIVE: If 2 angles are not congruent, then the angles are not right angles.

31. STATEMENT: If it is not raining, then I am happy.

CONVERSE: If I am happy, then it is not raining.

INVERSE: If it is raining, then I am not happy.

CONTRAPOSITIVE: If I am not happy, then it is raining.

32. Contrapositive

33. Given: $\overline{AB} \parallel \overline{CF}$

$\angle 2 \doteq \angle 3$

Prove: $\angle 1 \doteq \angle 3$

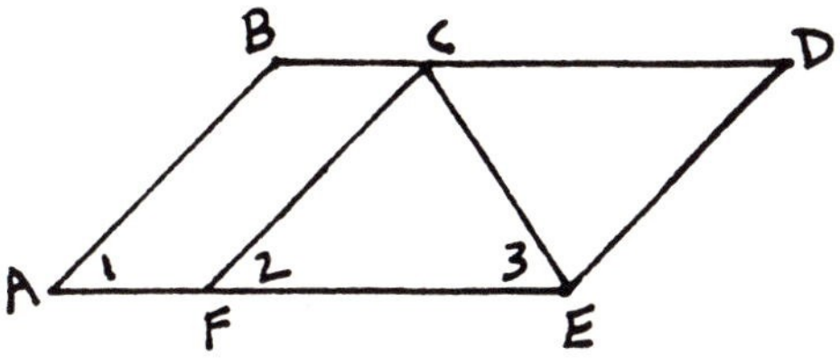

STATEMENTS	REASONS
1. $\overline{AB} \parallel \overline{CF}$	1. Given
2. $\angle 1 \doteq \angle 2$	2. If 2 $\parallel$ lines are cut by a trans., then corresponding $\angle$s are congruent.
3. $\angle 2 \doteq \angle 3$	3. Given
4. $\angle 1 \doteq \angle 3$	4. Transitive Prop. of Congruence

34. Given: $\angle 1$ is comp. to $\angle 2$

$\angle 2$ is comp. to $\angle 3$

Prove: $\overline{BD} \parallel \overline{AE}$

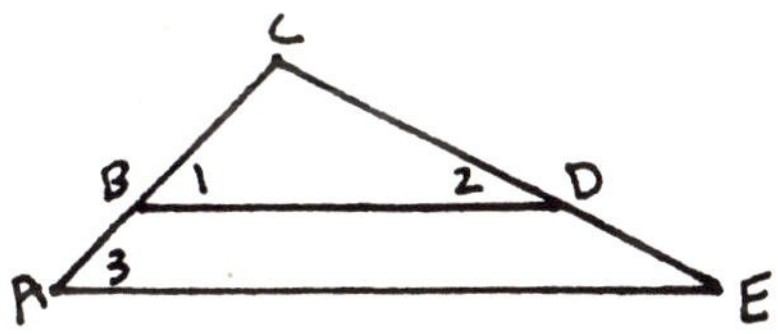

STATEMENTS	REASONS
1. $\angle 1$ is comp. to $\angle 2$	1. Given
2. $\angle 2$ is comp. to $\angle 3$	
3. $\angle 1 \doteq \angle 3$	2. If 2 $\angle$s are comp. to the same $\angle$, then these $\angle$s are $\doteq$.
4. $\overline{BD} \parallel \overline{AE}$	3. If 2 lines are cut by a trans. so that corresponding $\angle$s are $\doteq$, then the lines are $\parallel$.

35. Given: $\overline{BE} \perp \overline{DA}$

$\overline{CD} \perp \overline{DA}$

Prove: $\angle 1 \doteq \angle 2$

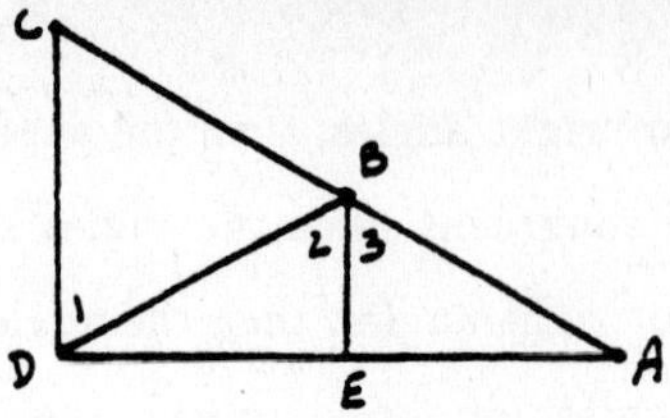

STATEMENTS	REASONS
1. $\overline{BE} \perp \overline{DA}$ $\overline{CD} \perp \overline{DA}$	1. Given
2. $\overline{BE} \parallel \overline{CD}$	2. If 2 lines are each $\perp$ to a 3rd line, then these lines are parallel to each other.
3. $\angle 1 \doteq \angle 2$	3. If 2 $\parallel$ lines are cut by a trans., then the alternate interior $\angle$s are $\doteq$.

36. Given: $\angle A \doteq \angle C$

$\overline{DC} \parallel \overrightarrow{AB}$

Prove: $\overline{DA} \parallel \overline{CB}$

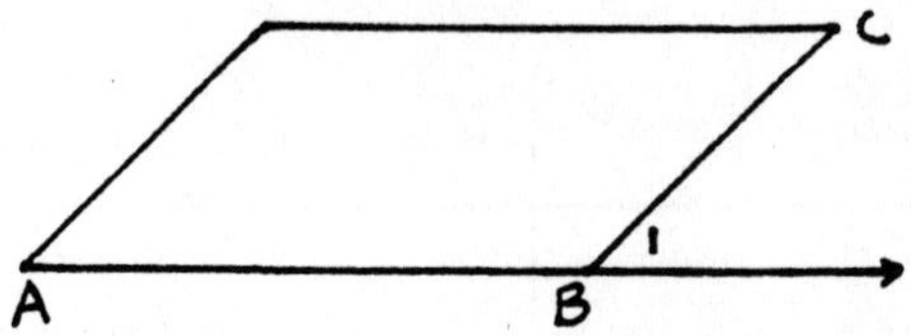

STATEMENTS	REASONS
1. $\angle A \doteq \angle C$	1. Given
2. $\overline{DC} \parallel \overrightarrow{AB}$	2. Given
3. $\angle C \doteq \angle 1$	3. If 2 $\parallel$ lines are cut by a trans., the alt. int. $\angle$s are congruent.
4. $\angle A \doteq \angle 1$	4. Transitive Prop. of Congruence
5. $\overline{DA} \parallel \overline{CB}$	5. If 2 lines are cut by a trans. so that corr. $\angle$s are $\doteq$, then these lines are $\parallel$.

37. Assume $x = -3$.

38. Assume the sides opposite these angles are $\doteq$.

39. Given: $m \nparallel n$

Prove: $\angle 1 \not\doteq \angle 2$

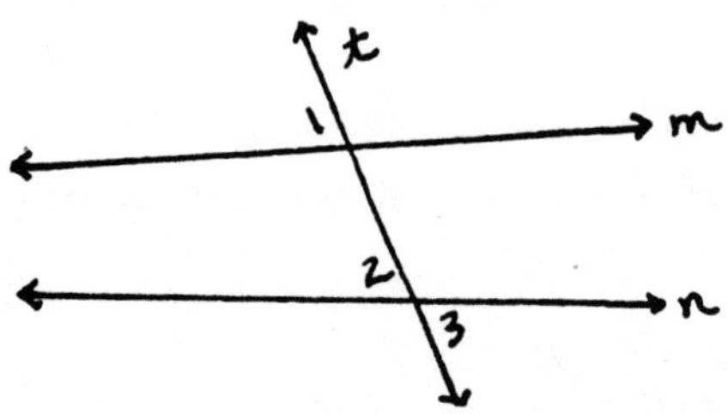

Indirect Proof:

Assume that $\angle 1 \doteq \angle 2$. Then $m \parallel n$ since congruent corr. angles are formed. But this contradicts our hypothesis. Therefore, our assumption must be false and $\angle 1 \not\doteq \angle 2$.

40. Given: $\angle 1 \not\doteq \angle 3$

Prove: m $\nparallel$ n

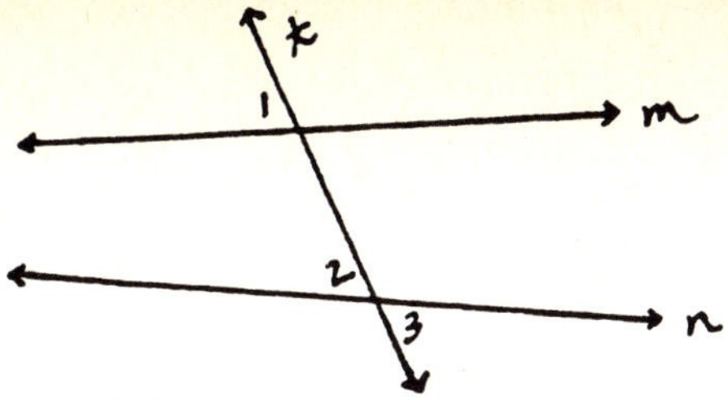

Indirect Proof:

Assume that m $\parallel$ n. Then $\angle 1 \doteq \angle 3$ since alt. ext. angles are congruent when parallel lines are cut by a transversal. But this contradicts the given fact that $\angle 1 \not\doteq \angle 3$. Therefore, our assumption must be false and it follows that m $\nparallel$ n.

41. Given: $\triangle$ ABC

Construct: The line through C parallel to $\overline{AB}$.

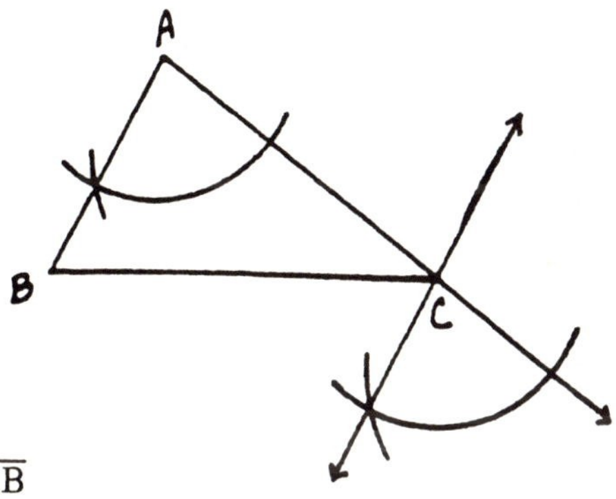

42. Given: $\overline{AB}$

Construct: An equilateral triangle ABC with side $\overline{AB}$.

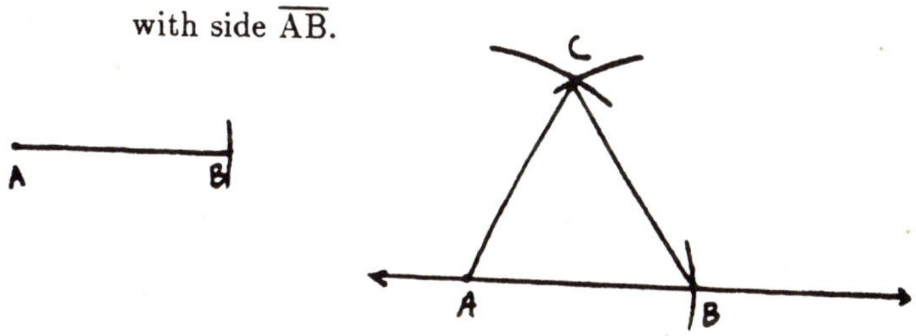

Chapter Three: Triangles

SECTION 3.1: Congruent Triangles

1. $\angle A$; $\overline{AB}$; No; No
5. SAS
9. SSS
13. ASA
17. SSS
21. $\overline{AD} \cong \overline{EC}$

25. 1. Given
    2. $\overline{AC} \cong \overline{AC}$
    3. SSS

29. Given: $\overline{AB} \perp \overline{BC}$ and
    $\overline{AB} \perp \overline{BD}$;
    also $\overline{BC} \cong \overline{BD}$
    Prove: $\triangle ABC \cong \triangle ABD$

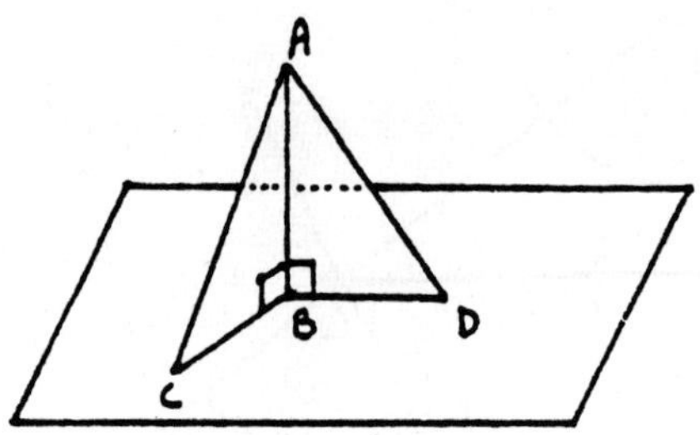

STATEMENTS	REASONS
1. $\overline{AB} \perp \overline{AC}$ and $\overline{AB} \perp \overline{BD}$	1. Given
2. $\angle ABC$ is a right $\angle$ and $\angle ABD$ is a right $\angle$	2. If 2 lines are $\perp$, then they meet to form a right $\angle$.
3. $\angle ABC \cong \angle ABD$	3. Any two right $\angle$s are $\cong$.
4. $\overline{BC} \cong \overline{BD}$	4. Given
5. $\overline{AB} \cong \overline{AB}$	5. Identity
6. $\triangle ABC \cong \triangle ABD$	6. SAS

33. Yes; SAS or SSS

37. a. $\triangle CBE$, $\triangle ADE$, $\triangle CDE$
    b. $\triangle ADC$
    c. $\triangle CBD$

1. Given: $\angle 1$ and $\angle 2$ are right $\angle$s

$\overline{CA} \doteq \overline{DA}$

Prove: $\triangle ABC \doteq \triangle ABD$

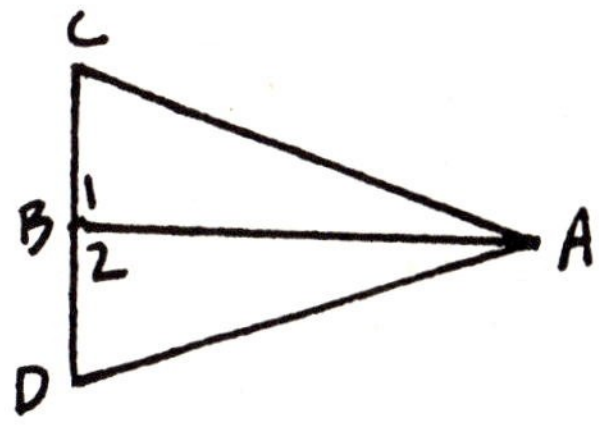

STATEMENTS	REASONS
1. $\angle 1$ and $\angle 2$ are right $\angle$s and $\overline{CA} \doteq \overline{DA}$	1. Given
2. $\overline{AB} \doteq \overline{AB}$	2. Identity
3. $\triangle ABC \doteq \triangle ABD$	3. HL

5. Given: $\angle R$ and $\angle V$ are right $\angle$s

$\angle 1 \doteq \angle 2$

Prove: $\triangle RST \doteq \triangle VST$

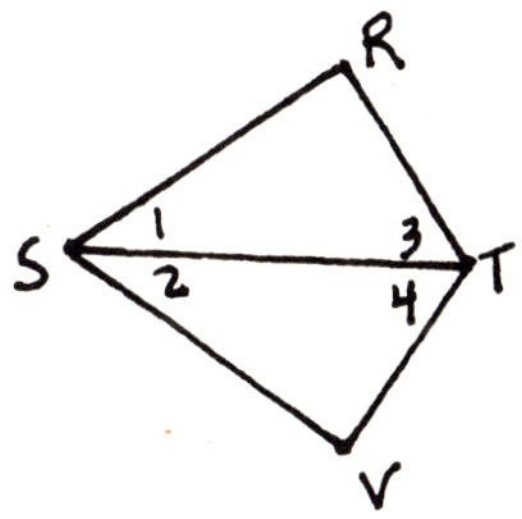

STATEMENTS	REASONS
1. $\angle R$ and $\angle V$ are right $\angle$s and $\angle 1 \doteq \angle 2$	1. Given
2. $\angle R \doteq \angle V$	2. All right $\angle$s are $\doteq$.
3. $\overline{ST} \doteq \overline{ST}$	3. Identity
4. $\triangle RST \doteq \triangle VST$	4. AAS

9. $m\angle 2 = 48$; $m\angle 3 = 48$;

$m\angle 5 = 42$; $m\angle 6 = 42$

13. Given: ∠s P and R are rt. ∠s

M is the midpoint of $\overline{PR}$

Prove: ∠N ≅ ∠Q

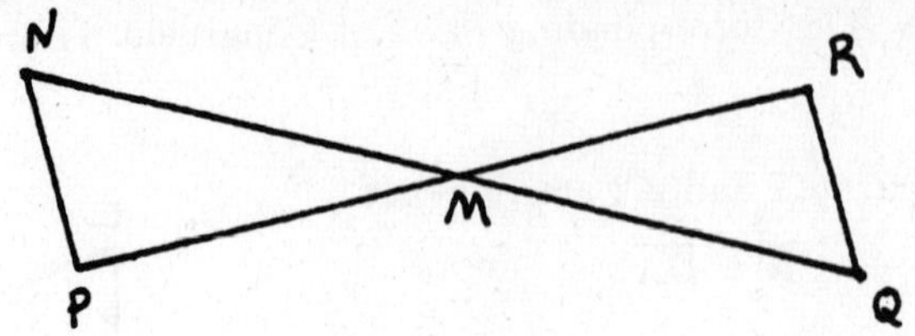

STATEMENTS	REASONS
1. ∠s P and R are rt. ∠s	1. Given
2. ∠P ≅ ∠R	2. All right ∠s are ≅.
3. M is the midpoint of $\overline{PR}$	3. Given
4. $\overline{PM} \cong \overline{MR}$	4. Midpoint of a segment forms 2 ≅ segments.
5. ∠NMP ≅ ∠QMR	5. If 2 lines intersect, the vertical angles formed are ≅
6. △NPM ≅ △QRM	6. ASA
7. ∠N ≅ ∠Q	7. CPCTC

17. c = 5

21. $c = \sqrt{41}$

25. Given: E is the midpoint of $\overline{FG}$;

$\overline{DF} \cong \overline{DG}$

Prove: $\overline{DE} \perp \overline{FG}$

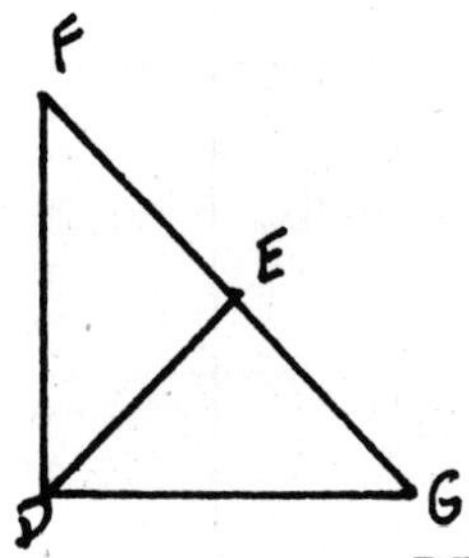

STATEMENTS	REASONS
1. E is the midpoint of $\overline{FG}$	1. Given
2. $\overline{FE} \cong \overline{EG}$	2. The midpoint of a segment forms 2 ≅ segments.
3. $\overline{DF} \cong \overline{DG}$	3. Given
4. $\overline{DE} \cong \overline{DE}$	4. Reflexive
5. △FDE ≅ △GDE	5. SSS
6. ∠DEF ≅ ∠DEG	6. CPCTC
7. $\overline{DE} \perp \overline{FG}$	7. If 2 lines meet to form ≅ adj. ∠s, then the lines are ⊥.

29. Given: $\overrightarrow{RW}$ bisects $\angle SRU$;

also $\overline{RS} \doteq \overline{RU}$

Prove: $\triangle TRU \doteq \triangle VRS$

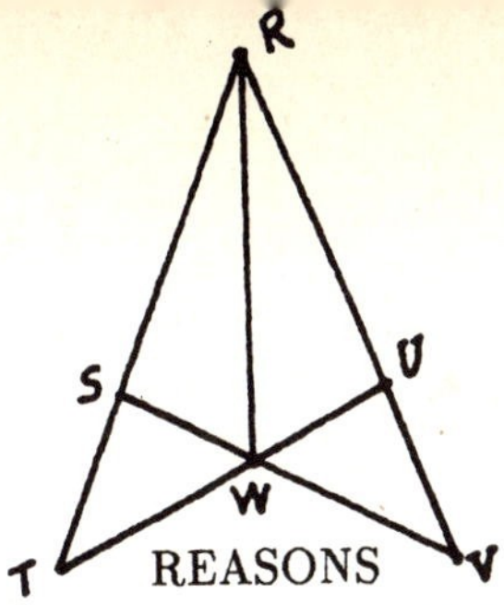

STATEMENTS	REASONS
1. $\overrightarrow{RW}$ bisects $\angle SRU$	1. Given
2. $\angle SRW \doteq \angle URW$	2. If a ray bisects an $\angle$, then 2 $\doteq$ $\angle$s are formed.
3. $\overline{RS} \doteq \overline{RU}$	3. Given
4. $\overline{RW} \doteq \overline{RW}$	4. identity
5. $\triangle RSW \doteq \triangle RUW$	5. SAS
6. $\angle RSW \doteq \angle RUW$	6. CPCTC
7. $\angle TRV \doteq \angle VRS$	7. Identity
8. $\triangle TRU \doteq \triangle VRS$	8. ASA

33. 751 feet

SECTION 3.3: Isosceles Triangles

1. Isosceles

5. $m\angle U = 69^\circ$

9. Underdetermined

13. Determined

17. $m\angle 2 = 68$ (Base $\angle$s in an isosceles $\triangle$ are $\doteq$)

Also, $m\angle 1 + m\angle 2 + m\angle 3 = 180$

$m\angle 1 + 68 + 68 = 180$

$m\angle 1 + 136 = 180$

$m\angle 1 = 44$

21. Let the measure of the vertex angle be x.

Then the measure of the base angles are

each x + 12.

$$x + (x + 12) + (x + 12) = 180$$

$$3x + 24 = 180$$

$$3x = 156$$

$$x = 52$$

$m\angle A = 52$; $m\angle B = 64$; $m\angle C = 64$.

25. 12

29. 1. Given

2. $\angle 3 \doteq \angle 2$

3. $\angle 1 \doteq \angle 2$

4. If 2 $\angle$s of one $\triangle$ are $\doteq$, then the opposite sides are $\doteq$.

33. Given: Isosceles $\triangle$ MNP with vertex P;
Isosceles $\triangle$ MNQ with vertex Q

Prove: $\triangle MQP \doteq \triangle NQP$

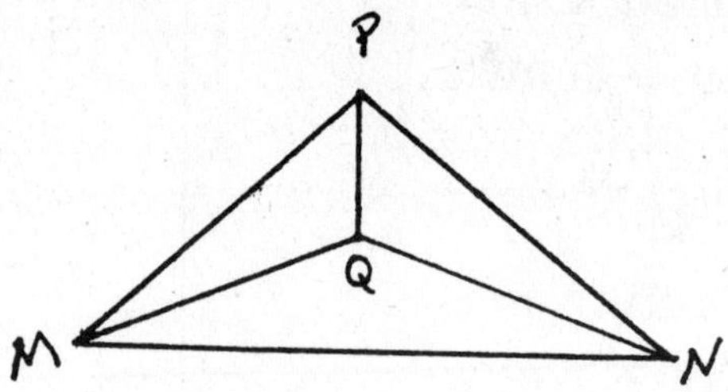

STATEMENTS	REASONS
1. Isosceles $\triangle$ MNP with vertex P	1. Given
2. $\overline{MP} \doteq \overline{NP}$	2. An isosceles $\triangle$ has 2 $\doteq$ sides.
3. Isosceles $\triangle$ MNQ with vertex Q	3. Given
4. $\overline{MQ} \doteq \overline{NQ}$	4. Same as (2)
5. $\overline{PQ} \doteq \overline{PQ}$	5. Identity
6. $\triangle MQP \doteq \triangle NQP$	6. SSS

37. 75

SECTION 3.4: Basic Constructions Justified

1.

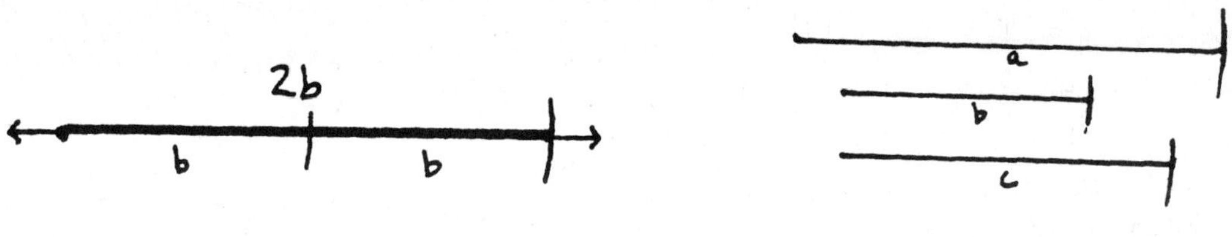

5.

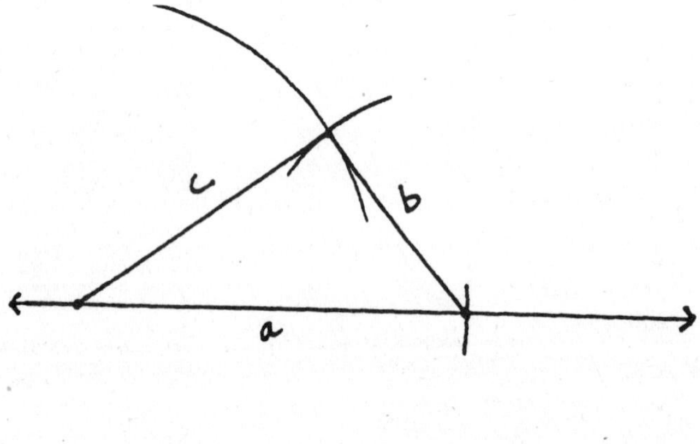

9.

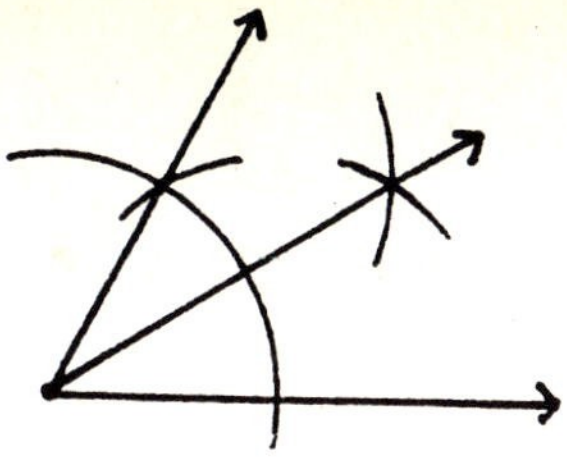

13.

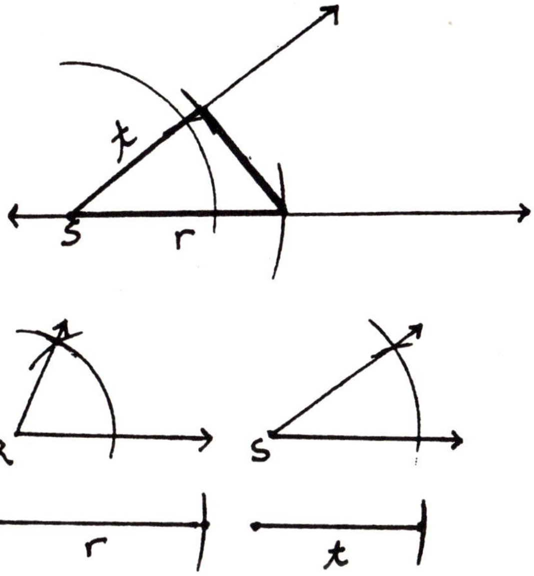

17.

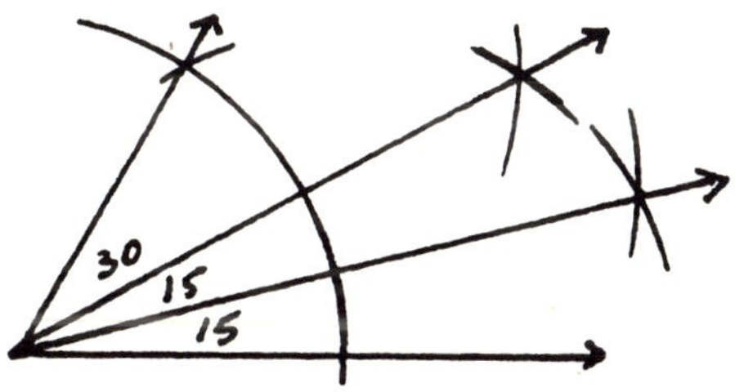

21.

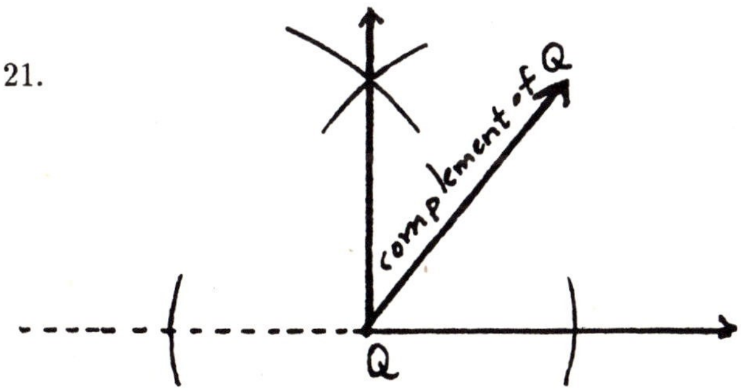

25.

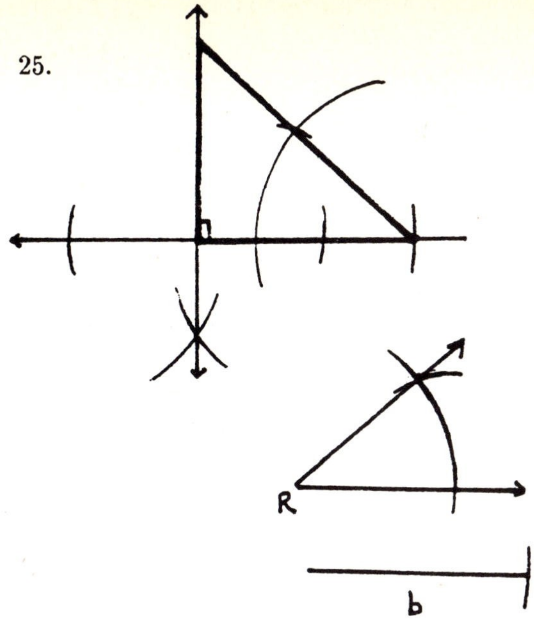

29. 120°

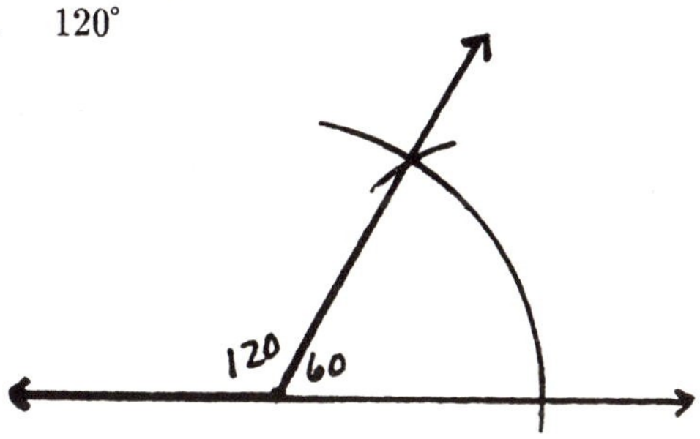

33. Yes

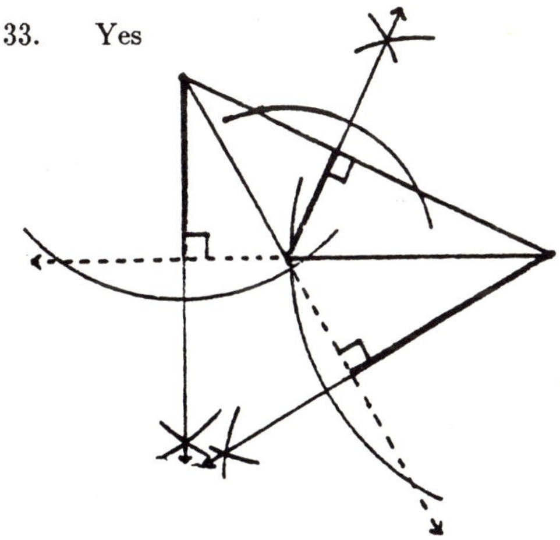

37. Point D is on the angle-bisector.

SECTION 3.5: Inequalities in a Triangle

1. False

5. True

9. True

13. a. Possible

    b. Not possible

    c. Not possible

17. Largest $\angle$ is 72° (two of these); smallest is 36°.

21. 1. $m\angle ABC > m\angle DBE$; $m\angle CBD > m\angle EBF$

    3. Angle-Addition Postulate

    4. $m\angle ABD > m\angle DBF$

25. $BC < EF$

29. $x + 2 < y < 5x + 12$

33. The length of an altitude of a triangle that does not contain a right angle is less than the length of either side containing the same vertex as the altitude.

Given: $\triangle ABC$ with altitude $\overline{BD}$ to $\overline{AC}$

Prove: $BD < AB$ and $BD < BC$

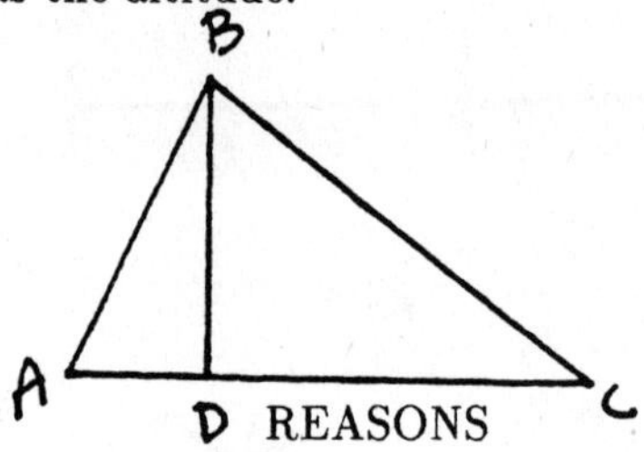

STATEMENTS	REASONS
1. $\triangle ABC$ with altitude $\overline{BD}$ to $\overline{AC}$	1. Given
2. $\overline{BD} \perp \overline{AC}$	2. Altitude of a $\triangle$ is the line segment drawn from a vertex $\perp$ to the opp. side.
3. $BD < AB$ and $BD < BC$	3. Shortest distance from a point to the line is the $\perp$ distance.

CHAPTER 3: Review

1. Given: $\angle AEB \doteq \angle DEC$

   $\overline{AE} \doteq \overline{ED}$

   Prove: $\triangle AEB \doteq \triangle DEC$

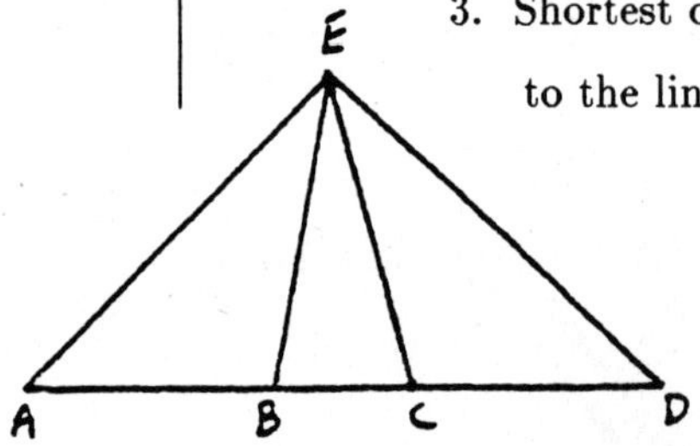

STATEMENTS	REASONS
1. $\angle AEB \doteq \angle DEC$	1. Given
2. $\overline{AE} \doteq \overline{ED}$	2. Given
3. $\angle A \doteq \angle D$	3. If 2 sides of a $\triangle$ are $\doteq$, then the $\angle$s opposite these sides are also $\doteq$.
4. $\triangle AEB \doteq \triangle DEC$	4. ASA

2. Given: $\overline{AB} \doteq \overline{EF}$

$\overline{AC} \doteq \overline{DF}$

$\angle 1 \doteq \angle 2$

Prove: $\angle B \doteq \angle E$

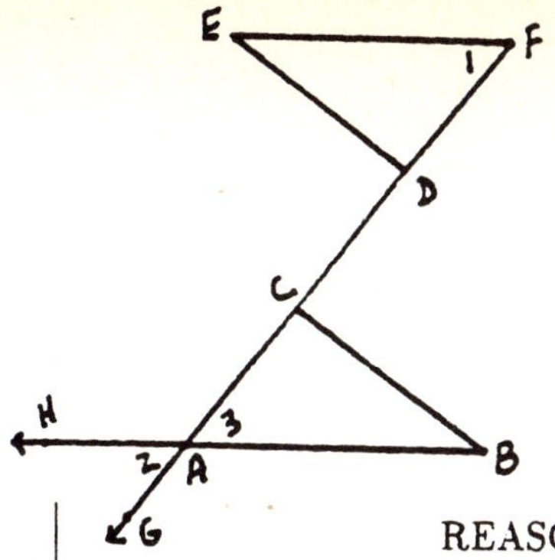

STATEMENTS	REASONS
1. $\overline{AB} \doteq \overline{EF}$	1. Given
2. $\overline{AC} \doteq \overline{DF}$; $\angle 1 \doteq \angle 2$	2. Given
3. $\angle 2 \doteq \angle 3$	3. If 2 lines intersect, then the vertical $\angle$s formed are $\doteq$.
4. $\angle 1 \doteq \angle 3$	4. Transitive Prop. for $\doteq$.
5. $\triangle ABC \doteq \triangle FED$	5. SAS
6. $\angle B \doteq \angle E$	6. CPCTC

Given: $\overline{AD}$ bisects $\overline{BC}$

$\overline{AB} \perp \overline{BC}$

$\overline{DC} \perp \overline{BC}$

Prove: $\overline{AE} \doteq \overline{ED}$

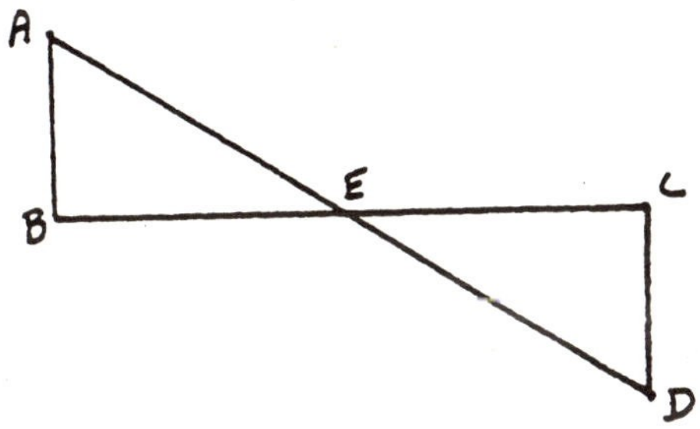

STATEMENTS	REASONS
1. $\overline{AD}$ bisects $\overline{BC}$	1. Given
2. $\overline{BE} \doteq \overline{EC}$	2. If a segment is bisected, 2 $\doteq$ segments are formed.
3. $\overline{AB} \perp \overline{BC}$ and $\overline{DC} \perp \overline{BC}$	3. Given
4. $\angle B$ is a rt. $\angle$ and $\angle C$ is a rt. $\angle$	4. If 2 lines are $\perp$, they meet to form a rt. $\angle$.
5. $\angle B \doteq \angle C$	5. Any 2 right $\angle$s are $\doteq$.
6. $\angle AEB \doteq \angle DEC$	6. If 2 lines intersect, then the vertical $\angle$s formed are $\doteq$.
7. $\triangle ABE \doteq \triangle DCE$	7. ASA
8. $\overline{AE} \doteq \overline{ED}$	8. CPCTC

4. Given: $\overline{OA} \doteq \overline{OB}$

$\overline{OC}$ is the median to $\overline{AB}$

Prove: $\overline{OC} \perp \overline{AB}$

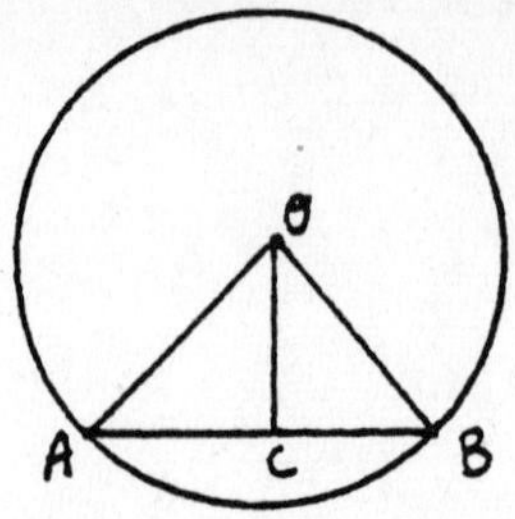

STATEMENTS	REASONS
1. $\overline{OA} \doteq \overline{OB}$	1. Given
2. $\overline{OC}$ is the median to $\overline{AB}$	2. Given
3. C is the midpoint of $\overline{AB}$	3. The median of a $\triangle$ is a segment drawn from a vertex to the midpoint of the opp. side.
4. $\overline{AC} \doteq \overline{CB}$	4. Midpoint of seg. forms 2 $\doteq$ segments.
5. $\overline{OC} \doteq \overline{OC}$	5. Identity
6. $\triangle AOC \doteq \triangle BOC$	6. SSS
7. $\angle OCA \doteq \angle OCB$	7. CPCTC
8. $\overline{OC} \perp \overline{AB}$	8. If 2 lines meet to form $\doteq$ adj. $\angle$s, then the lines are $\perp$.

5. Given: $\overline{AB} \doteq \overline{DE}$

$\overline{AB} \parallel \overline{DE}$

$\overline{AC} \doteq \overline{DF}$

Prove: $\overline{BC} \parallel \overline{FE}$

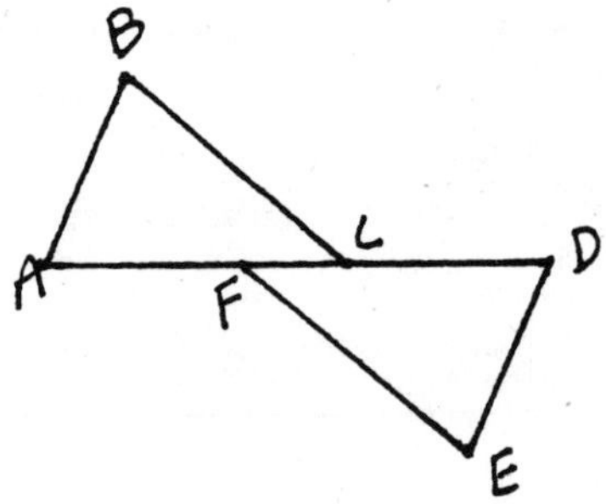

STATEMENTS	REASONS
1. $\overline{AB} \doteq \overline{DE}$ and $\overline{AB} \parallel \overline{DE}$	1. Given
2. $\angle A \doteq \angle D$	2. If 2 $\parallel$ lines are cut by a trans., then the alt. int. $\angle$s are $\doteq$.
3. $\overline{AC} \doteq \overline{DF}$	3. Given
4. $\triangle BAC \doteq \triangle EDF$	4. SAS
5. $\angle BCA \doteq \angle EFD$	5. CPCTC
6. $\overline{BC} \parallel \overline{FE}$	6. If 2 lines are cut by a trans. so that alt. int. $\angle$s are $\doteq$, then the lines are $\parallel$.

6. Given: B is the midpoint of $\overline{AC}$

$\overline{BD} \perp \overline{AC}$

Prove: $\triangle ADC$ is isosceles

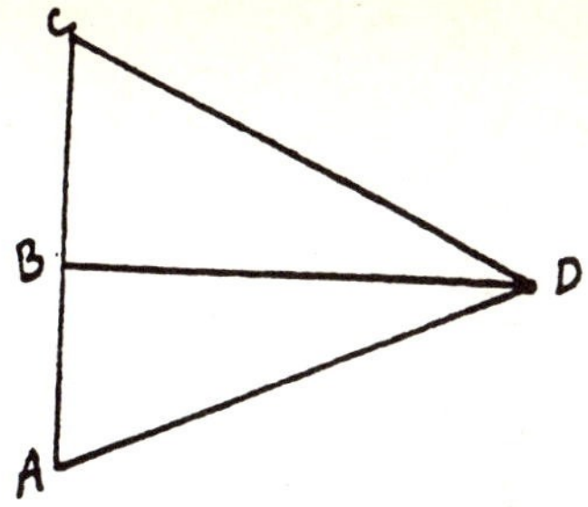

STATEMENTS	REASONS
1. B is the midpoint of $\overline{AC}$	1. Given
2. $\overline{CB} \doteq \overline{BA}$	2. Midpoint of a segment forms 2 $\doteq$ segments.
3. $\overline{BD} \perp \overline{AC}$	3. Given
4. $\angle DBC \doteq \angle DBA$	4. If 2 lines are $\perp$, they meet to form $\doteq$ adj. $\angle$s.
5. $\overline{BD} \doteq \overline{BD}$	5. Identity
6. $\triangle CBD \doteq \triangle ABD$	6. SAS
7. $\overline{DC} \doteq \overline{DA}$	7. CPCTC
8. $\triangle ADC$ is isosceles	8. If a $\triangle$ has 2 $\doteq$ sides, it is an isos. $\triangle$.

7. Given: $\overline{JM} \perp \overline{GM}$

$\overline{GK} \perp \overline{KJ}$

$\overline{GH} \doteq \overline{HJ}$

Prove: $\overline{GM} \doteq \overline{JK}$

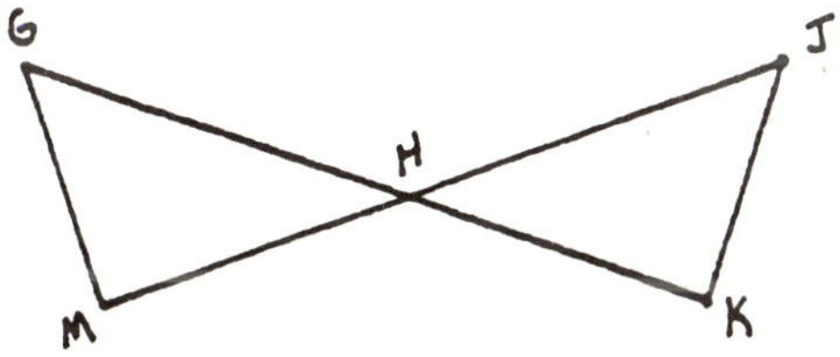

STATEMENTS	REASONS
1. $\overline{JM} \perp \overline{GM}$ and $\overline{GK} \perp \overline{KJ}$	1. Given
2. $\angle M$ is a rt. $\angle$ and $\angle K$ is a rt. $\angle$	2. If 2 lines are $\perp$, they meet to form a rt. $\angle$.
3. $\angle M \doteq \angle K$	3. Any 2 rt. $\angle$s are $\doteq$.
4. $\overline{GH} \doteq \overline{HJ}$	4. Given
5. $\angle GHM \doteq \angle JHK$	5. If 2 lines intersect, the vertical $\angle$s formed are $\doteq$.
6. $\triangle GHM \doteq \triangle JHK$	6. AAS
7. $\overline{GM} \doteq \overline{JK}$	7. CPCTC

8. Given: $\overline{TN} \doteq \overline{TR}$

$\overline{TO} \perp \overline{NP}$ and

$\overline{TS} \perp \overline{PR}$

$\overline{TO} \doteq \overline{TS}$

Prove: $\angle N \doteq \angle R$

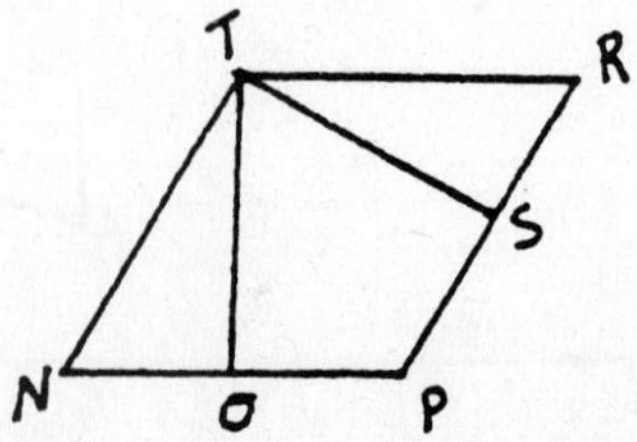

STATEMENTS	REASONS
1. $\overline{TN} \doteq \overline{TR}$	1. Given
2. $\overline{TO} \perp \overline{NP}$ and $\overline{TS} \perp \overline{PR}$	2. Given
3. $\angle TON$ is a rt. $\angle$ and $\angle TSR$ is a rt. $\angle$.	3. If 2 lines are $\perp$, they meet to form a rt. $\angle$.
4. $\overline{TO} \doteq \overline{TS}$	4. Given
5. $\triangle TON \doteq \triangle TSR$	5. HL
6. $\angle N \doteq \angle R$	6. CPCTC

9. Given: $\overline{YZ}$ is the base of an isosceles triangle

$\overrightarrow{XA} \parallel \overline{YZ}$

Prove: $\angle 1 \doteq \angle 2$

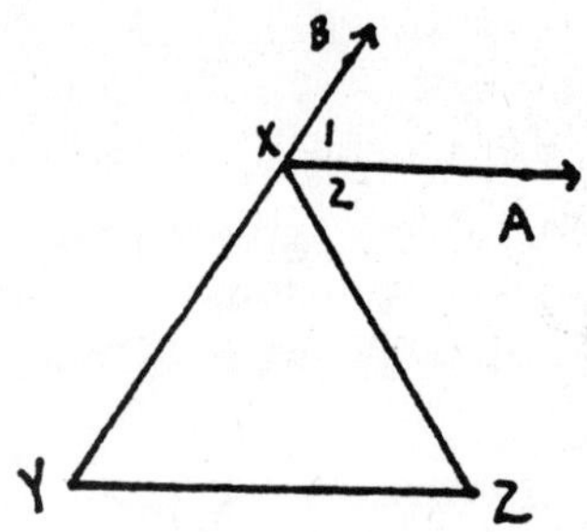

STATEMENTS	REASONS
1. $\overline{YZ}$ is the base of an isosceles triangle	1. Given
2. $\angle Y \doteq \angle Z$	2. Base $\angle$s of an isos. $\triangle$ are $\doteq$.
3. $\overrightarrow{XA} \parallel \overline{YZ}$	3. Given
4. $\angle 1 \doteq \angle Y$	4. If 2 $\parallel$ lines are cut by a trans., then the corresponding $\angle$s are $\doteq$.
5. $\angle 2 \doteq \angle Z$	5. If 2 $\parallel$ lines are cut by a trans., then the alt. int. $\angle$s are $\doteq$.
6. $\angle 1 \doteq \angle 2$	6. Transitive Prop. for $\doteq$.

10. Given: $\overline{AB} \parallel \overline{DC}$

$\overline{AB} \doteq \overline{DC}$

C is the midpoint of $\overline{BE}$

Prove: $\overline{AC} \parallel \overline{DE}$

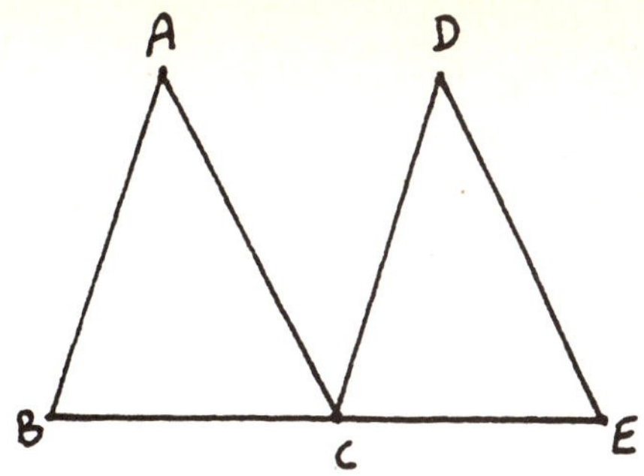

STATEMENTS	REASONS
1. $\overline{AB} \parallel \overline{DC}$	1. Given
2. $\angle B \doteq \angle DCE$	2. If 2 $\parallel$ lines are cut by a trans., then the corresp. $\angle$s are $\doteq$.
3. $\overline{AB} \doteq \overline{DC}$	3. Given
4. C is the midpoint of $\overline{BE}$	4. Given
5. $\overline{BC} \doteq \overline{CE}$	5. Midpoint of a seg. forms 2 $\doteq$ segments.
6. $\triangle ABC \doteq \triangle DCE$	6. SAS
7. $\angle ACB \doteq \angle E$	7. CPCTC
8. $\overline{AC} \parallel \overline{DE}$	8. If 2 lines are cut by a trans. so that the corresp. $\angle$s are $\doteq$, then the lines are $\parallel$.

11. Given: $\angle BAD \doteq \angle CDA$

$\overline{AB} \doteq \overline{CD}$

Prove: $\overline{AE} \doteq \overline{ED}$

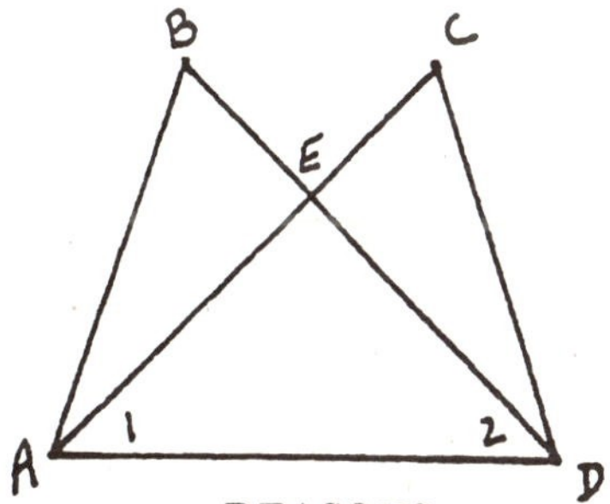

STATEMENTS	REASONS
1. $\angle BAD \doteq \angle CDA$	1. Given
2. $\overline{AB} \doteq \overline{CD}$	2. Given
3. $\overline{AD} \doteq \overline{AD}$	3. Identity
4. $\triangle BAD \doteq \triangle CDA$	4. SAS
5. $\angle 1 \doteq \angle 2$	5. CPCTC
6. $\overline{AE} \doteq \overline{ED}$	6. If 2 $\angle$s of a triangle are $\doteq$, then the sides opp. these $\angle$s are also $\doteq$.

12. Given: $\overline{BE}$ is altitude to $\overline{AC}$

$\overline{AD}$ is altitude to $\overline{CE}$

$\overline{BC} \doteq \overline{CD}$

Prove: $\overline{BE} \doteq \overline{AD}$

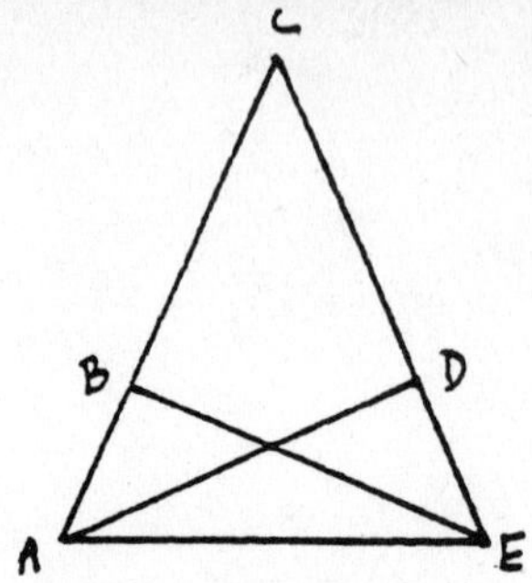

STATEMENTS	REASONS
1. $\overline{BE}$ is altitude to $\overline{AC}$ and $\overline{AD}$ is altitude to $\overline{CE}$	1. Given
2. $\overline{BE} \perp \overline{AC}$ and $\overline{AD} \perp \overline{CE}$	2. An altitude is a line segment drawn from a vertex $\perp$ to opp. side.
3. $\angle CBE$ is a rt. $\angle$ and $\angle CDA$ is a rt. $\angle$	3. If 2 lines are $\perp$, they meet to form a rt. $\angle$.
4. $\angle CBE \doteq \angle CDA$	4. Any 2 rt. $\angle$s are $\doteq$.
5. $\overline{BC} \doteq \overline{CD}$	5. Given
6. $\angle C \doteq \angle C$	6. Identity
7. $\triangle CBE \doteq \triangle CDA$	7. ASA
8. $\overline{BE} \doteq \overline{AD}$	8. CPCTC

13. Given: $\overline{AB} \doteq \overline{CD}$

$\angle BAD \doteq \angle CDA$

Prove: $\triangle AED$ is isosceles

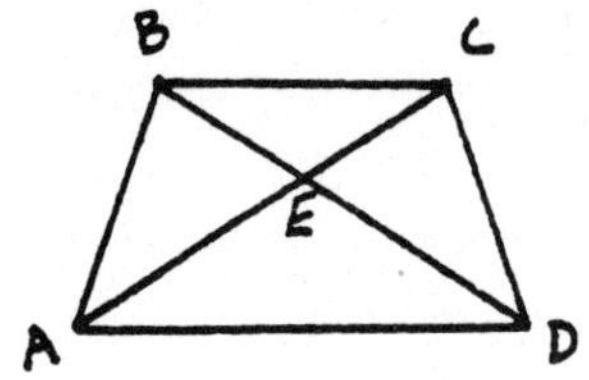

STATEMENTS	REASONS
1. $\overline{AB} \doteq \overline{CD}$	1. Given
2. $\angle BAD \doteq \angle CDA$	2. Given
3. $\overline{AD} \doteq \overline{AD}$	3. Identity
4. $\triangle BAD \doteq \triangle CDA$	4. SAS
5. $\angle CAD \doteq \angle BDA$	5. CPCTC
6. $\overline{AE} \doteq \overline{ED}$	6. If 2 $\angle$s of a $\triangle$ are $\doteq$, then the sides opp. these $\angle$s are also $\doteq$.
7. $\triangle AED$ is isosceles	7. If a $\triangle$ has 2 $\doteq$ sides, it is an isos. $\triangle$.

14. Given: $\overrightarrow{AC}$ bisects $\angle BAD$

Prove: $AD > CD$

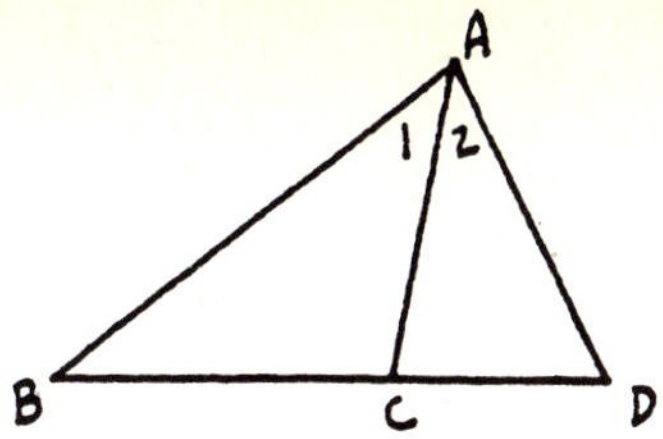

STATEMENTS	REASONS
1. $\overrightarrow{AC}$ bisects $\angle BAD$	1. Given
2. $m\angle 1 = m\angle 2$	2. If a ray bisects an $\angle$, it forms 2 $\angle$s of = measure.
3. $m\angle ACD > m\angle 1$	3. The measure of an ext. $\angle$ of a $\triangle$ is greater than the measure of either of the nonadjacent interior angles.
4. $m\angle ACD > m\angle 2$	4. Substitution
5. $AD > CD$	5. If the measure of one angle of a $\triangle$ is greater than the measure of a second angle, then the side which is opposite the 1st angle is longer than the side which is opp. the second angle.

15. a. $\overline{PR}$
    b. $\overline{PQ}$

16. $\overline{BC}$, $\overline{AC}$, $\overline{AB}$

17. $\angle R$, $\angle Q$, $\angle P$

18. $\overline{AD}$

19. (b)

20. 5 and 35

21. 20

22. 115

23. $3x + 10 = \frac{5}{2}x + 18$

$\frac{1}{2}x = 8$

$x = 16$

$m\angle 4 = \frac{5}{2}(16) + 18 = 58$; $m\angle C = 64$

24. $10 + x + 6 + 2x - 3 = 40$

$3x + 13 = 40$

$3x = 27$

$x = 9$

$AB = 10$; $BC = 15$; $AC = 15$; the triangle is isosceles.

25. Either $AB = BC$ or $AB = AC$ or $BC = AC$

If $AB = BC$, $y + 7 = 3y + 5$

$-2y = -2$

$y = 1$

If $AB = AC$, $y + 7 = 9 - y$

$2y = -2$

$y = 1$

If $BC = AC$, $3y + 5 = 9 - y$

$4y = 4$

$y = 1$

If $y = 1$, $AB = 8$; $BC = 8$; $AC = 8$; the triangle is also equilateral.

26. If $m\angle 1 = 5x$, then the

$m\angle 2 = 180 - 5x$

$m\angle 4 = m\angle 2 = 180 - 5x$.

But $\angle 3 = m\angle 4$,

therefore $2x + 12 = 180 - 5x$

$7x = 168$

$x = 24$;

$m\angle 2 = 180 - 5(24) = 60$.

27. Construct an angle that measures 75°.

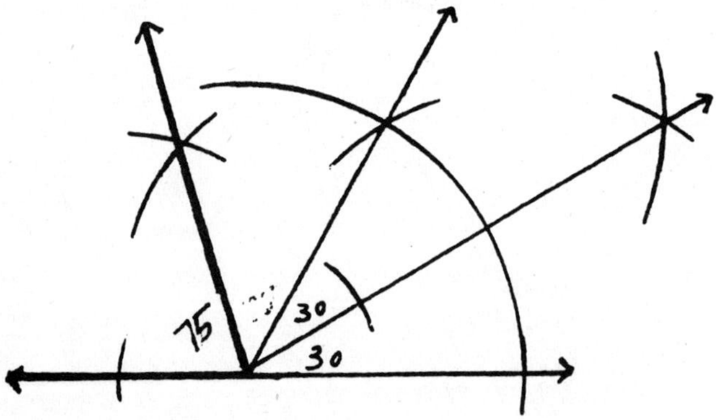

28. Construct a right triangle that has acute angle A and hypotenuse of length c.

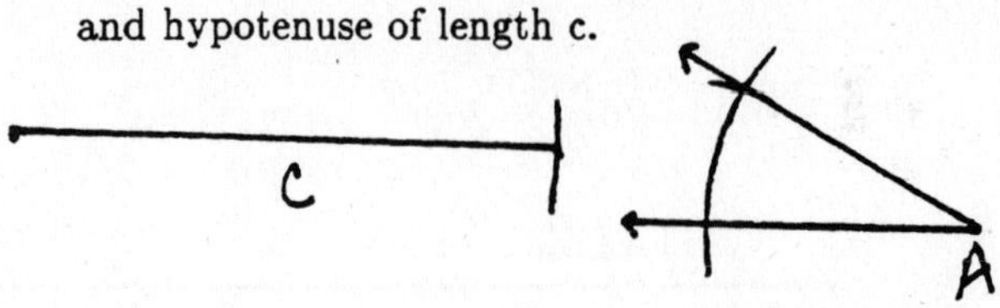

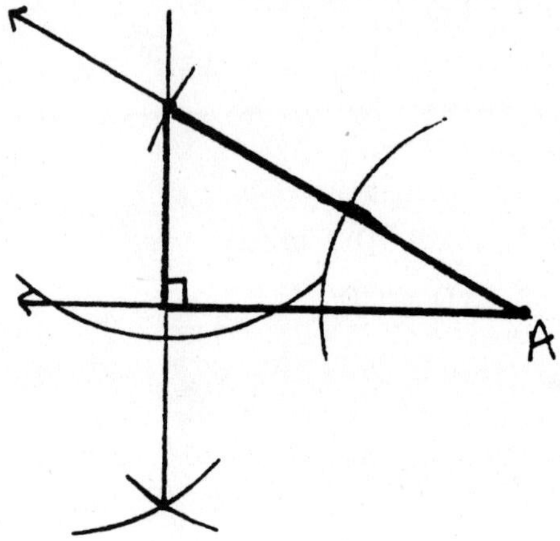

29. Construct another isosceles triangle in which the base angles are half as large as the given base angles.

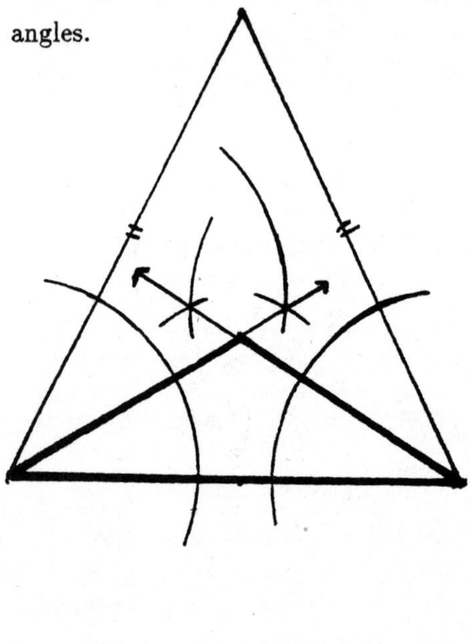

Chapter Four: Quadrilaterals

SECTION 4.1: Properties of Parallelograms

1. a. AB = DC

   b. AD = BC

5. In parallelogram ABCD, AB = DC.

   Therefore,

   $3x + 2 = 5x - 2$

   $4 = 2x$

   $x = 2$

   AB = DC = 8; BC = AD = 9.

9. $\overline{AC}$

13. The resulting quadrilateral appears to be a parallelogram.

17. 1. Given
    2. $\overline{RV} \perp \overline{VT}$ and $\overline{ST} \perp \overline{VT}$
    3. $\overline{RV} \parallel \overline{ST}$
    4. RSTV is a parallelogram

21. The opposite angles of a parallelogram are congruent.

    Given: Parallelogram ABCD

    Prove: $\angle BAD \doteq \angle BCD$ and

    $\angle ABC \doteq \angle ADC$

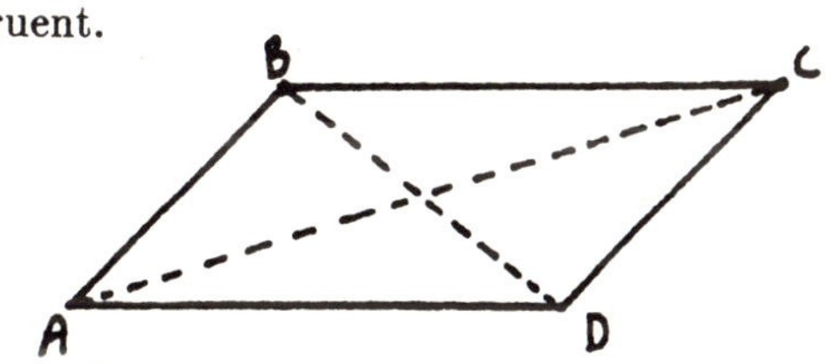

STATEMENTS	REASONS
1. Parallelogram ABCD	1. Given
2. Draw diagonal $\overline{BD}$	2. Through 2 points there is exactly one line.
3. $\triangle ABD \doteq \triangle CDB$	3. A diagonal of a parallelogram separates it into 2 $\doteq$ $\triangle$s.
4. $\angle BAD \doteq \angle BCD$	4. CPCTC
5. Draw in diagonal $\overline{AC}$	5. Same as (2)
6. $\triangle ABC \doteq \triangle CDA$	6. Same as (3)
7. $\angle ABC \doteq \angle ADC$	7. CPCTC

25. $\angle P$ is a right angle.

29. 255 mph

SECTION 4.2: The Parallelogram and Kite

1. a. Yes
   b. No

5. a. Kite
   b. Parallelogram

9. 6.18

13. 10

17. 1. Given
    2. Identity
    3. $\triangle NMQ \cong \triangle NPQ$
    4. CPCTC
    5. MNPQ is a kite.

21. Given: Kite HJKL with diagonal $\overline{HK}$

Prove: $\overrightarrow{HK}$ bisects $\angle LHJ$

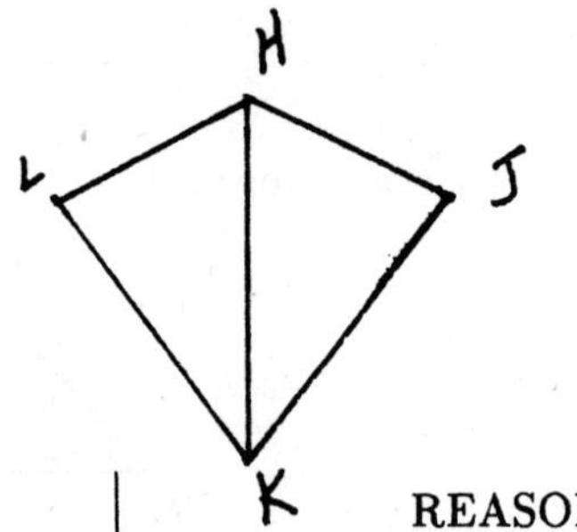

STATEMENTS	REASONS
1. Kite HJKL with diagonal $\overline{HK}$	1. Given
2. $\overline{LH} \cong \overline{HJ}$ and $\overline{LK} \cong \overline{JK}$	2. A kite is a quad. with 2 distinct pairs of $\cong$ adjacent sides.
3. $\overline{HK} \cong \overline{HK}$	3. Identity
4. $\triangle LHK \cong \triangle JHK$	4. SSS
5. $\angle LHK \cong \angle JHK$	5. CPCTC
6. $\overrightarrow{HK}$ bisects $\angle LHJ$	6. If a ray divides an $\angle$ into 2 $\cong$ $\angle$s, then the ray bisects the $\angle$.

25. In a kite, one diagonal is the perpendicular bisector of the other diagonal.

Given: Kite ABCD with diagoanls $\overline{AC}$ and $\overline{BD}$

Prove: $\overline{BD} \perp \overline{AC}$ and $\overline{BD}$ bisects $\overline{AC}$

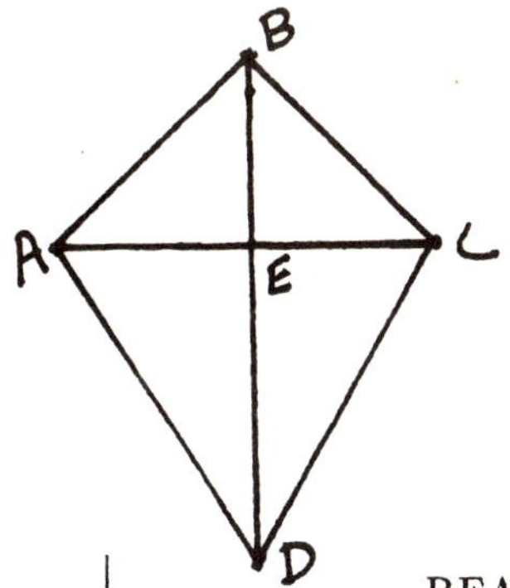

STATEMENTS	REASONS
1. Kite ABCD with diagonals $\overline{AC}$ and $\overline{BD}$	1. Given
2. $\overline{AB} \doteq \overline{BC}$ and $\overline{AD} \doteq \overline{CD}$	2. A kite is a quad. with 2 distinct pairs of $\doteq$ adjacent sides.
3. $\overline{BD} \doteq \overline{BD}$	3. Identity
4. $\triangle ABD \doteq \triangle CBD$	4. SSS
5. $\angle ABD \doteq \angle CBD$	5. CPCTC
6. $\overline{BE} \doteq \overline{BE}$	6. Identity
7. $\triangle ABE \doteq \triangle CBE$	7. SAS
8. $\angle BEA \doteq \angle BEC$	8. CPCTC
9. $\overline{BD} \perp \overline{AC}$	9. If 2 lines intersect to form $\doteq$ adjacent $\angle$s, then the lines are $\perp$.
10. $\overline{AE} \doteq \overline{EC}$	10. CPCTC
11. $\overline{BD}$ bisects $\overline{AC}$	11. If a segment divides another segment into 2 $\doteq$ segments, then the segment is bisected.

29. If M and N are the midpoints of $\overline{RS}$ and $\overline{RT}$ and if RM = RN, then RS = RT = 4x + 2. But if m$\angle$R = 60 and RS = RT, then m$\angle$S = m$\angle$T = 60. Triangle RST is equiangular and therefore equilateral. Therefore 4x + 2 = 5x − 3 or x = 5. If x = 5, RM = 11 and ST = 22.

SECTION 4.3: The Rectangle, Square, and Rhombus

1. m$\angle$A = 60; m$\angle$ABC = 120

5. The quadrilateral is a rhombus.

9. 2x + 7 = 3x + 2

   x = 5; AD = BC = 3(5) + 4 = 19

13. QP = $\sqrt{72} = 6\sqrt{2}$; MN = $\sqrt{72} = 6\sqrt{2}$

17. AD = $\sqrt{34}$

21. 1. Given

    4. The line joining the midpoints of two sides of a $\triangle$ is $\parallel$ to the third side.

    5. If 2 lines are each $\parallel$ to a third line, then the 2 lines are $\parallel$.

    6. Same as (2)

    7. Same as (3)

    8. Same as (4)

    9. Same as (5)

    10. ABCD is a parallelogram

25. A square can be defined as a rhombus with a right angle. If all the sides of a rhombus are congruent (#16), then all the sides of a square must be congruent since the square is a rhombus.

29. If the diagonals of a parallelogram are perpendicular, then the parallelogram is a rhombus.

Given: ABCD is a parallelogram with $\overline{BD} \perp \overline{AC}$.

Prove: ABCD is a rhombus

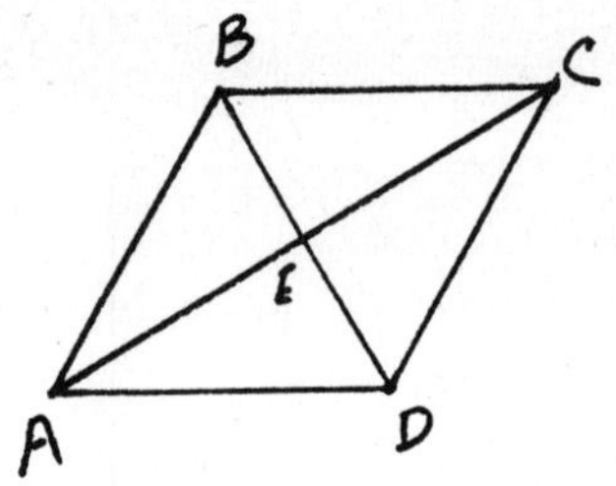

STATEMENTS	REASONS
1. ABCD is a parallelogram with $\overline{BD} \perp \overline{AC}$	1. Given
2. $\angle BEA \doteq \angle BEC$	2. If 2 lines are $\perp$, then they meet to form $\doteq$ adj. $\angle$s.
3. $\overline{BD}$ bisects $\overline{AC}$	3. Diagonals of a parallelogram bisect each other.
4. $\overline{AE} \doteq \overline{EC}$	4. If a seg. is bisected, then 2 $\doteq$ segments are formed.
5. $\overline{BE} \doteq \overline{BE}$	5. Identity
6. $\triangle BEA \doteq \triangle BEC$	6. SAS
7. $\overline{AB} \doteq \overline{BC}$	7. CPCTC
8. ABCD is a rhombus	8. If a parallelogram has 2 adj. sides $\doteq$, then the parallelogram is a rhombus.

33. Let the length of the ramp = x.

$x^2 = 4^2 + 20^2$

$x^2 = 16 + 400$

$x^2 = 416$

$x = 20.4$

SECTION 4.4: The Trapezoid

1. $m\angle D = 180 - 58 = 122$

   $m\angle B = 180 - 125 = 55$

5. The quadrilateral is a rhombus.

9. $MN = \frac{1}{2}(AB + DC)$

   $9.5 = \frac{1}{2}(8.2 + DC)$

   $19 = 8.2 + DC$

   $DC = 10.8$

13. Given: ABCD is an isosceles trapezoid

    Prove: $\triangle ABE$ is isosceles

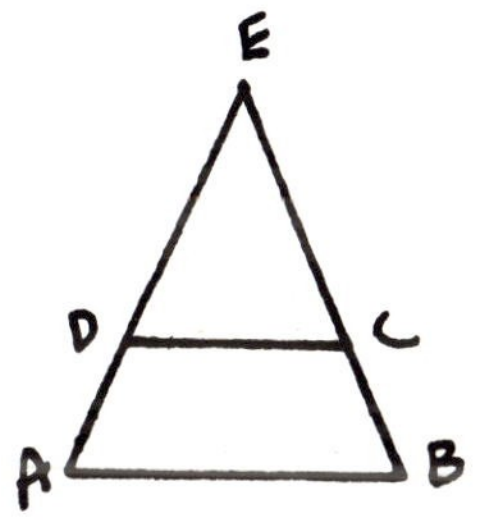

STATEMENTS	REASONS
1. ABCD is an isosceles trap.	1. Given
2. $\angle A \cong \angle B$	2. Lower base angles of an isosceles trap. are $\cong$.
3. $\overline{EB} \cong \overline{EA}$	3. If 2 $\angle$s of a $\triangle$ are $\cong$, then the sides opposite these $\angle$s are also $\cong$.
4. $\triangle ABE$ is isosceles	4. If a $\triangle$ has 2 $\cong$ sides, it is an isosceles $\triangle$.

17. $(QP)^2 = (MQ)^2 + (MP)^2$

$13^2 = 5^2 + (MP)^2$

$169 = 25 + (MP)^2$

$144 = (MP)^2$

$12 = MP$

21. $AB = BC$

$2x + 3 = x + 7$

$x = 4$

$EF = DE = 3x + 2$

$EF = 3(4) + 2 = 14$

25. If 2 consecutive angles of a quadrilateral are supplementary, the quadrilateral is a trapezoid.

Given: Quadrilateral ABCD with $\angle A$ is supp. to $\angle B$

Prove: ABCD is a trapezoid

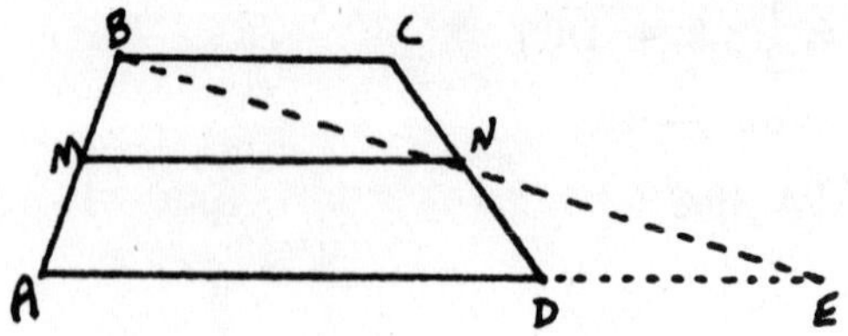

STATEMENTS	REASONS
1. Quadrilateral ABCD with $\angle A$ is supp. to $\angle B$	1. Given
2. $\overline{BC} \parallel \overline{AD}$	2. If 2 lines are cut by a trans. so that the interior $\angle$s on the same side of the trans. are supp., then these lines are $\parallel$.
3. ABCD is a trapezoid	3. If a quad. had 2 $\parallel$ sides, then the quad. is a trapezoid.

29. Given: $\overline{EF}$ is the median of trapezoid ABCD

Prove: $EF = \frac{1}{2}(AB + DC)$

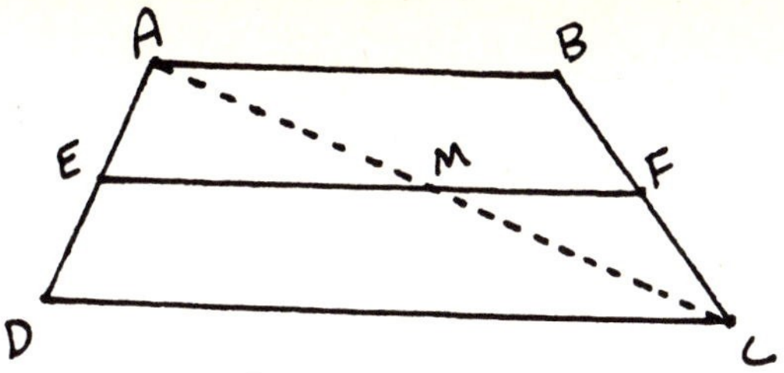

STATEMENTS	REASONS
1. $\overline{EF}$ is the median of trapezoid ABCD	1. Given
2. $\overline{AB} \parallel \overline{DC}$	2. Trapezoid has one pair of $\parallel$ sides.
3. $\overline{EF} \parallel$ to both $\overline{AB}$ and $\overline{DC}$	3. The median of a trap. is $\parallel$ to each base.
4. E is the midpoint of $\overline{AD}$ and F is the midpoint of $\overline{BC}$	4. The median of a trap. joins the midpoints of the nonparallel sides.
5. $\overline{AE} \cong \overline{ED}$	5. The midpoint of a segment forms 2 $\cong$ segments.
6. $\overline{AM} \cong \overline{MC}$	6. If 3 (or more) parallel lines intercept $\cong$ segments on one transversal, then they intercept $\cong$ segments on any transversal.
7. M is the midpoint of $\overline{AC}$	7. If a point divides a segment into 2 $\cong$ segments, then the point is the midpoint.
8. In $\triangle ADC$, $EM = \frac{1}{2}(DC)$ and in $\triangle ABC$, $MF = \frac{1}{2}(AB)$	8. The segment that joins the midpoints of two sides of a $\triangle$ is $\parallel$ to the third side and has a length equal to one-half the length of the third side.
9. $EM + MF = \frac{1}{2}(AB) + \frac{1}{2}(DC)$	9. Addition Property of Equality
10. $EM + MF = \frac{1}{2}(AB + DC)$	10. Distributive Property
11. $EF = EM + MF$	11. Segment Addition Postulate
12. $EF = \frac{1}{2}(AB + DC)$	12. Substitution

Chapter 4: Review

1. A

2. S

3. N

4. S

5. S

6. A

7. A

8. A

9. A

10. N

11. S

12. N

13. $2(2x + 3) + 2(5x - 4) = 96$

$4x + 6 + 10x - 8 = 96$

$14x - 2 = 96$

$14x = 98$

$x = 7$

$AB = DC = 2(7) + 3 = 17$

$AD = BC = 5(7) - 4 = 31$

14. $2x + 6 + x + 24 = 180$

$3x + 30 = 180$

$3x = 150$

$x = 50$

$m\angle C = m\angle A = 2(50) + 6 = 106$

15. The sides of the parallelogram measure 13 since $5^2 + 12^2 = (\text{side})^2$. Perimeter is 52.

16. $4x = 2x + 50$

$2x = 50$

$x = 25$

$m\angle M = 4(25) = 100$; $m\angle P = 180 - 100 = 80$

17. $\overline{PN}$

18. Kite

19. $m\angle G = m\angle F = 180 - 108 = 72$

$m\angle E = 108$

20. Median $= \frac{1}{2}(12.3 + 17.5)$

$= \frac{1}{2}(29.8)$

14.9

21. $15 = \frac{1}{2}(3x + 2 + 2x - 7)$

$30 = 5x - 5$

$35 = 5x$

$x = 7$

$MN = 3(7) + 2 = 23$

$PO = 2(7) - 7 = 7$

22. If $\overline{FJ} \cong \overline{FH}$ and M and N are their midpoints, then $FM = NH$ or $2y + 3 = 5y - 9$

$-3y = -12$

$y = 4$

$FM = 2(4) + 3 = 11$;

$FN = NH = 5(4) - 9 = 11$;

$JH = 2(4) = 8$. The perimeter of $\triangle FMN = 26$.

23. Since M and N are midpoints, $\overline{MN} \parallel \overline{JH}$ and $MN = \frac{1}{2} \cdot JH$. Therefore, $MN = 6$, $m\angle FMN = 80$ and $m\angle FNM = 40$.

24. Since M and N are midpoints, $MN = \frac{1}{2} \cdot JH$.

Therefore $x^2 + 6 = \frac{1}{2} \cdot 2x(x + 2)$

$x^2 + 6 = x(x + 2)$

$x^2 + 6 = x^2 + 2x$

$6 = 2x$

$x = 3$;

$MN = 15$; $JH = 30$.

25. Given: ABCD is a parallelogram

$\overline{AF} \doteq \overline{CE}$

Prove: $\overline{DF} \parallel \overline{EB}$

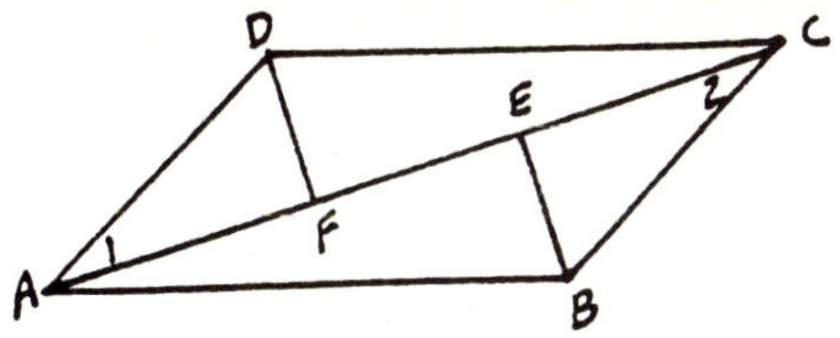

STATEMENTS	REASONS
1. ABCD is a parallelogram	1. Given
2. $\overline{AD} \doteq \overline{CB}$	2. Opp. sides of a parallelogram are $\doteq$.
3. $\overline{AD} \parallel \overline{CB}$	3. Opp. sides of a parallelogram are $\parallel$.
4. $\angle 1 \doteq \angle 2$	4. If 2 $\parallel$ lines are cut by a trans., then the alternate interior $\angle$s are $\doteq$.
5. $\overline{AF} \doteq \overline{CE}$	5. Given
6. $\triangle DAF \doteq \triangle BCE$	6. SAS
7. $\angle DFA \doteq \angle BEC$	7. CPCTC
8. $\overline{DF} \parallel \overline{EB}$	8. If 2 lines are cut by a trans. so that alt. ext. $\angle$s are $\doteq$, then the lines are $\parallel$.

26. Given: ABEF is a rect.

BCDE is a rect.

$\overline{FE} \doteq \overline{ED}$

Prove: $\overline{AE} \doteq \overline{BD}$ and $\overline{AE} \parallel \overline{BD}$

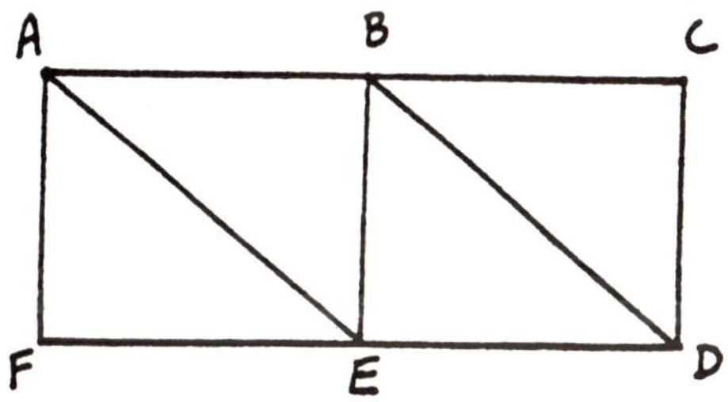

STATEMENTS	REASONS
1. ABEF is a rect.	1. Given
2. ABEF is a parallelogram	2. A rect. is a parallelogram with a rt. $\angle$.
3. $\overline{AF} \doteq \overline{BE}$	3. Opp. sides of a parallelogram are $\doteq$.
4. BCDE is a rect.	4. Given
5. $\angle F$ and $\angle BED$ are rt. $\angle$s	5. Same as (2)
6. $\angle F \doteq \angle BED$	6. Any 2 rt. $\angle$s are $\doteq$.
7. $\overline{FE} \doteq \overline{ED}$	7. Given
8. $\triangle AFE \doteq \triangle BED$	8. SAS
9. $\overline{AE} \doteq \overline{BD}$	9. CPCTC
10. $\angle AEF \doteq \angle BDE$	10. CPCTC
11. $\overline{AE} \parallel \overline{BD}$	11. If lines are cut by a trans. so that the corresp. $\angle$s are $\doteq$, then the lines are $\parallel$.

27. Given: $\overline{DE}$ is a median in $\triangle ADC$

$\overline{BE} \doteq \overline{FD}$

$\overline{EF} \doteq \overline{FD}$

Prove: ABCF is a parallelogram

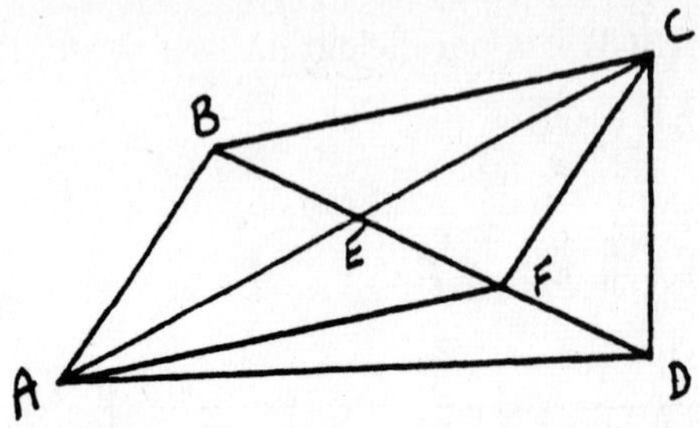

STATEMENTS	REASONS
1. $\overline{DE}$ is a median in $\triangle ADC$	1. Given
2. E is the midpoint of $\overline{AC}$	2. Median of a $\triangle$ is a line segment drawn from a vertex to the midpoint of the opp. side.
3. $\overline{AE} \doteq \overline{EC}$	3. Midpoint of a segment forms 2 = segments.
4. $\overline{BE} \doteq \overline{FD}$ and $\overline{EF} \doteq \overline{FD}$	4. Given
5. $\overline{BE} \doteq \overline{EF}$	5. Transitive Prop. for $\doteq$
6. ABCF is a parallelogram	6. If the diagonals of a quad. bisect each other, then the quad. is a parallelogram.

28. Given: $\triangle FAB \doteq \triangle HCD$

$\triangle EAD \doteq \triangle GCB$

Prove: ABCD is a parallelogram

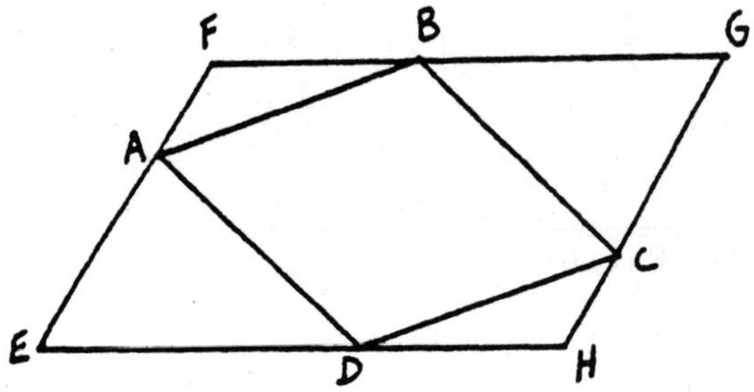

STATEMENTS	REASONS
1. $\triangle FAB \doteq \triangle HCD$	1. Given
2. $\overline{AB} \doteq \overline{DC}$	2. CPCTC
3. $\triangle EAD \doteq \triangle GCB$	3. Given
4. $\overline{AD} \doteq \overline{BC}$	4. CPCTC
5. ABCD is a parallelogram	5. If a quad. has both pairs of opp. sides $\doteq$, then the quad. is a parallelogram.

29. Given: ABCD is a parallelogram

$\overline{DC} \doteq \overline{BN}$

$\angle 3 \doteq \angle 4$

Prove: ABCD is a rhombus

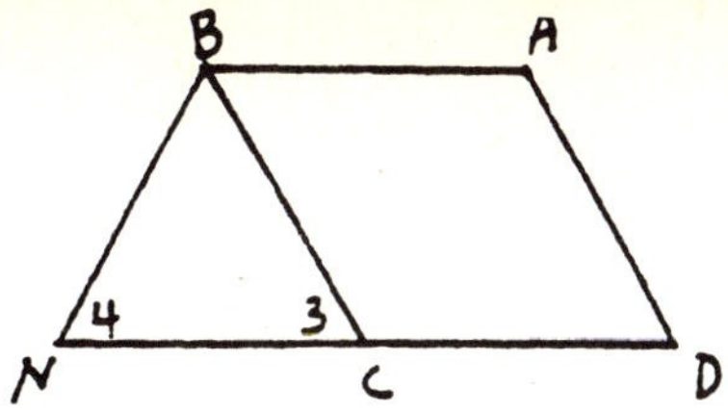

STATEMENTS	REASONS
1. ABCD is a parallelogram	1. Given
2. $\overline{DC} \doteq \overline{BN}$	2. Given
3. $\angle 3 \doteq \angle 4$	3. Given
4. $\overline{BN} \doteq \overline{BC}$	4. If 2 $\angle$s of a $\triangle$ are $\doteq$, then the sides opp. these $\angle$s are also $\doteq$.
5. $\overline{DC} \doteq \overline{BC}$	5. Transitive Prop. for $\doteq$.
6. ABCD is a rhombus	6. If a parallelogram has 2 $\doteq$ adj. sides, then the parallelogram is a rhombus.

30. Given: $\triangle$TWX is isosceles with base $\overline{WX}$

$\overline{RY} \parallel \overline{WX}$

Prove: RWXY is an isosceles trapezoid

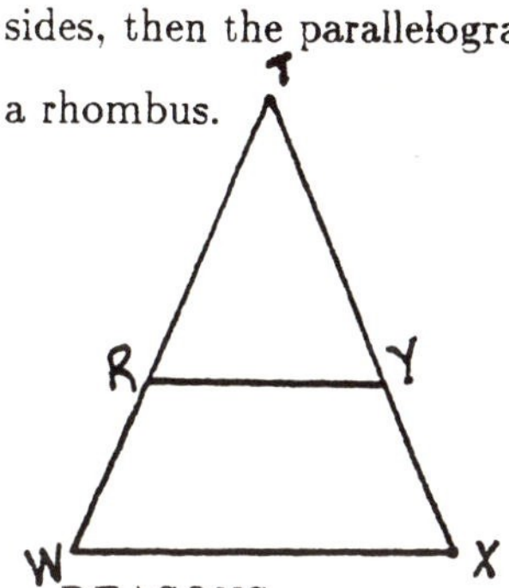

STATEMENTS	REASONS
1. $\triangle$TWX is isosceles with base $\overline{WX}$	1. Given
2. $\angle W \doteq \angle X$	2. Base $\angle$s of an isos. $\triangle$ are $\doteq$.
3. $\overline{RY} \parallel \overline{WX}$	3. Given
4. $\angle TRY \doteq \angle W$ and $\angle TYR \doteq \angle X$	4. If 2 $\parallel$ lines are cut by a trans., then the corresp. $\angle$s are $\doteq$.
5. $\angle TRY \doteq \angle TYR$	5. Transitive Prop. for $\doteq$.
6. $\overline{TR} \doteq \overline{TY}$	6. If 2 $\angle$s of a $\triangle$ are $\doteq$, then the side opp. these $\angle$s are also $\doteq$.
7. $\overline{TW} \doteq \overline{TX}$	7. Isosceles $\triangle$ has 2 $\doteq$ sides.
8. TR = TY and TW = TX	8. If 2 segments are $\doteq$, then they are equal in length.
9. TW = TR + RW and TX = TY + YX	9. Segment-Addition Post.
10. TR + RW = TY + YX	10. Substitution
11. RW = YX	11. Subtraction Prop. of Eq.

12. $\overline{RW} \doteq \overline{YX}$	12. If segments are = in length, then they are $\doteq$.
13. RWXY is an isosceles trapezoid	13. If a quad. has one pair of ∥ sides and the non-parallel sides are $\doteq$, then the quad. is an isos. trap.

31. Construct a rhombus given these lengths for the diagonals.

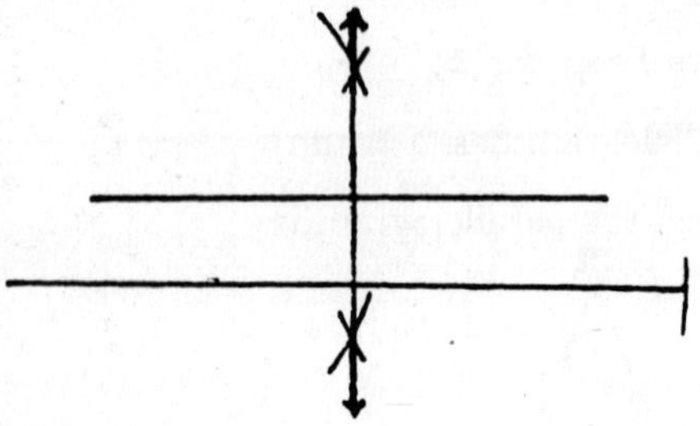

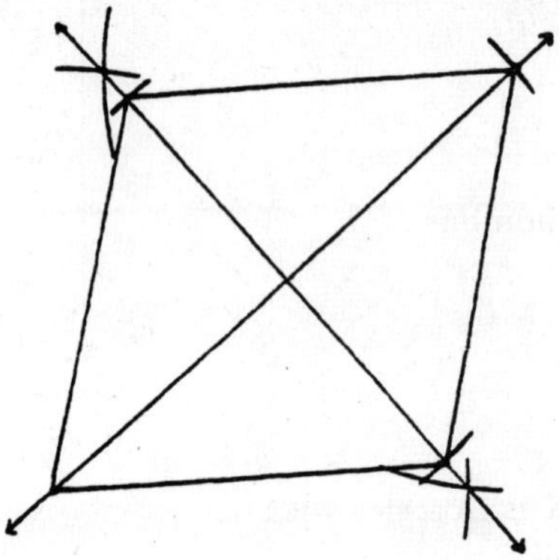

32. a. $\overline{AB} \perp \overline{BC}$

 b. AC = 13

33. a. $\overline{WY} \perp \overline{XZ}$

 b. WY = 30

Chapter Five: Similar Triangles

SECTION 5.1: Ratios, Rates, and Proportions

1. a. $\frac{12}{15} = \frac{4}{5}$

   b. $\frac{12 \text{ inches}}{15 \text{ inches}} = \frac{4}{5}$

   c. $\frac{1 \text{ foot}}{18 \text{ inches}} = \frac{12 \text{ inches}}{18 \text{ inches}} = \frac{2}{3}$

   d. $\frac{1 \text{ foot}}{18 \text{ ounces}}$ is incommensurable

5. a. $12x = 36$

   $x = 3$

   b. $21x = 168$

   $x = 8$

9. a. $x^2 = 28$

   $x = \pm\sqrt{28} = \pm\sqrt{47} = \pm 2\sqrt{7} \approx 5.29$

   b. $x^2 = 18$

   $x = \pm\sqrt{18} = \pm\sqrt{9 \cdot 2} = \pm 3\sqrt{2} \approx 4.24$

13. a. $3(x + 1) = 2x^2$

   $3x + 3 = 2x^2$

   $2x^2 - 3x - 3 = 0$

   $a = 2 \quad b = -3 \quad c = -3$

   $$x = \frac{-b \pm \sqrt{b^2 + -4ac}}{2a}$$

   $$x = \frac{3 \pm \sqrt{9 + -4(2)(-3)}}{2(2)}$$

   $$x = \frac{3 \pm \sqrt{9 + 24}}{4}$$

   $$x = \frac{3 \pm \sqrt{33}}{4} \approx 2.19 \text{ or } -0.69$$

   b. $5(x + 1) = 2x(x - 1)$

   $5x + 5 = 2x^2 - 2x$

   $0 = 2x^2 - 7x - 5$

   $a = 2 \quad b = -7 \quad c = -5$

   $$x = \frac{-b \pm \sqrt{b^2 + -4ac}}{2a}$$

   $$x = \frac{7 \pm \sqrt{49 + -4(2)(-5)}}{2(2)}$$

   $$x = \frac{7 \pm \sqrt{49 + 40}}{4}$$

   $$x = \frac{7 \pm \sqrt{89}}{4} \approx 4.19 \text{ or } -0.61$$

17. $\frac{4 \text{ eggs}}{3 \text{ cups of milk}} = \frac{14 \text{ eggs}}{x \text{ cups of milk}}$

   $4x = 42$

   $x = \frac{42}{4}$ or $10\frac{1}{2}$ cups of milk

21. a. $\frac{BD}{AD} = \frac{AD}{DC}$

   $\frac{6}{AD} = \frac{AD}{8}$

   $(AD)^2 = 48$

   $AD = \sqrt{48} = \sqrt{16 \cdot 3} = 4\sqrt{3} \approx 6.93$

   b. $\frac{BD}{AD} = \frac{AD}{DC}$

   $\frac{BD}{6} = \frac{6}{8}$

   $8(BD) = 36$

   $BD = \frac{36}{8} = 4\frac{1}{2}$

25. Let the first angle have measure x so that the complementary angle has measure $90 - x$. Then

$$\frac{x}{90 - x} = \frac{4}{5}$$

$$5x = 4(90 - x)$$

$$5x = 360 - 4x$$

$$9x = 360$$

$$x = 40;\ 90 - x = 50.$$

The angles measure 40 and 50.

Alternate method: Let the measures of the two angles be $4x$ and $5x$. Then

$$4x + 5x = 90$$

$$9x = 90$$

$$x = 10;\ 4x = 40;\ 5x = 50.$$

The angles measure 40 and 50.

29. $\frac{7}{3} = \frac{6}{YZ}$

$$7 \cdot YZ = 18$$

$$YZ = 2\tfrac{4}{7} \approx 2.57$$

33. $\frac{1 \text{ in}}{3 \text{ ft}} = \frac{x \text{ in}}{12 \text{ ft}}$

$$3x = 12$$

$$x = 4$$

$$\frac{1 \text{ in}}{3 \text{ ft}} = \frac{y \text{ in}}{14 \text{ ft}}$$

$$3y = 14$$

$$y = 4\tfrac{2}{3} \text{ in}$$

The blue print should be 4 in by $4\frac{2}{3}$ in.

SECTION 5.2: Similar Triangles

1 . a. $\triangle ABC \sim \triangle XTN$

b. $\triangle ACB \sim \triangle NXT$

5. SSS $\sim$

9. SAS ~

13. a. Yes

b. Yes

c. Yes

17. 1. ∠H ≐ ∠F

2. If 2 ∠s are vertical ∠s, then they are ≐.

3. △HJK ~ △FGK; AA

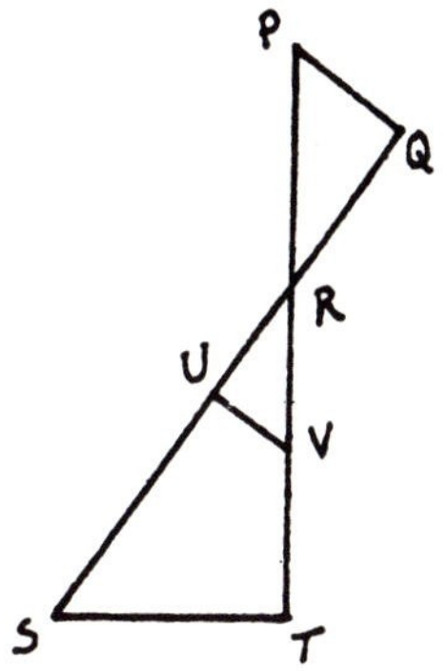

21. Given: ∠RVU ≐ ∠S and ∠RUV ≐ ∠Q

Prove: △PQR ~ △STR

Proof: By hypothesis, ∠RVU ≐ ∠S and ∠RUV ≐ ∠Q. ∠PRQ ≐ ∠URV since they are vertical angles. If 2 ∠s in one △ are ≐ to 2 ∠s in another △, then the third ∠s are also ≐. Therefore, ∠P ≐ ∠UVR. By the Transitive Property, ∠P ≐ ∠S. Thus, △PQR ~ △STR by AA.

25. Prove: The lengths of the corresponding altitudes of similar triangles have the same ratio as the lengths of any pair of corresponding sides.

Given: △DEF ~ △MNP

$\overline{DG}$ and $\overline{MQ}$ are altitudes

Prove: $\frac{DG}{MQ} = \frac{DE}{MN}$

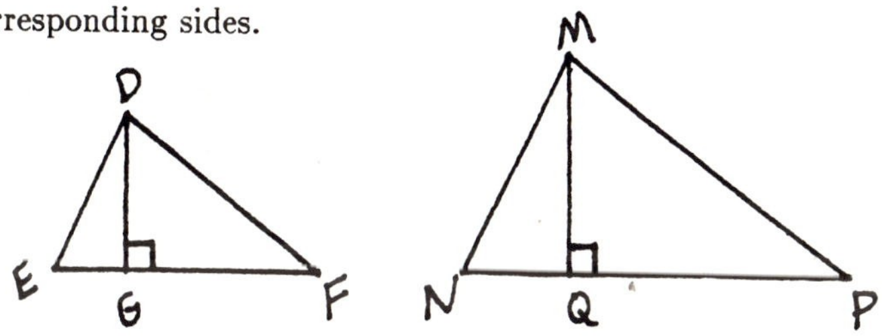

STATEMENTS	REASONS
1. △DEF ~ △MNP $\overline{DG}$ and $\overline{MQ}$ are altitudes	1. Given
2. $\overline{DG} \perp \overline{EF}$ and $\overline{MQ} \perp \overline{NP}$	2. An altitude is a segment drawn from a vertex ⊥ to the opposite side.
3. ∠DGE and ∠MQN are rt. ∠s	3. ⊥ lines form a rt. ∠
4. ∠DGE ≐ ∠MQN	4. Right ∠s are ≐.
5. ∠E ≐ ∠N	5. If 2 △s are ~, then the corresponding ∠s are ≐.
6. △DGE ~ △MQN	6. AA
7. $\frac{DG}{MQ} = \frac{DE}{MN}$	7. Corresponding sides of ~ △s are proportional.

29. Let DB = x; then AB = x + 4

$\frac{x}{8} + \frac{x+4}{10}$

$10x = 8(x + 4)$

$10x = 8x + 32$

$2x = 32$

$x = 16$; DB = 16

33. Let BC = x = CE; then

$AE = 20 - x$

$\frac{x}{4} = \frac{20}{20 - x}$

$x(20 - x) = 80$

$20x - x^2 = 80$

$x^2 - 20x + 80 = 0$

$a = 1 \quad b = -20 \quad c = 80$

$$x = \frac{-b \pm \sqrt{b^2 + -4ac}}{2a}$$

$$x = \frac{20 \pm \sqrt{400 + -4(1)(80)}}{2(1)}$$

$$x = \frac{20 \pm \sqrt{400 + -320}}{2}$$

$$x = \frac{20 \pm \sqrt{80}}{2}$$

$$x = \frac{20 \pm \sqrt{16 \cdot 5}}{2}$$

$$x = \frac{20 \pm 4\sqrt{5}}{2}$$

$x = 10 \pm 2\sqrt{5}$; BC = $10 + 2\sqrt{5}$ or $10 - 2\sqrt{5} \approx 14.47$ or 5.53

37. Let the height of the tree be x.

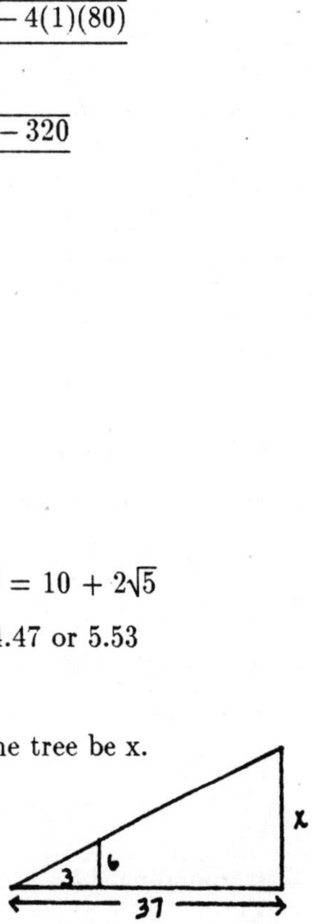

$\frac{3}{6} = \frac{37}{x}$

$\frac{1}{2} = \frac{37}{x}$

$x = 74$

The height of the tree is 74 feet.

41. The altitude drawn to the hypotenuse of a right triangle separates the right triangle into two right triangles which are similar to each other and to the original triangle.

Given: Right $\triangle ABC$ with the right $\angle$ at ACB; $\overline{CD}$ is the altitude to $\overline{AB}$.

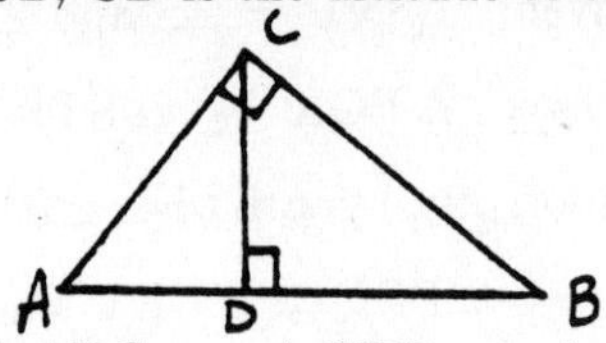

Prove: $\triangle ADC \sim \triangle CDB \sim \triangle ACD$

Proof: Draw the three $\triangle$s separately.

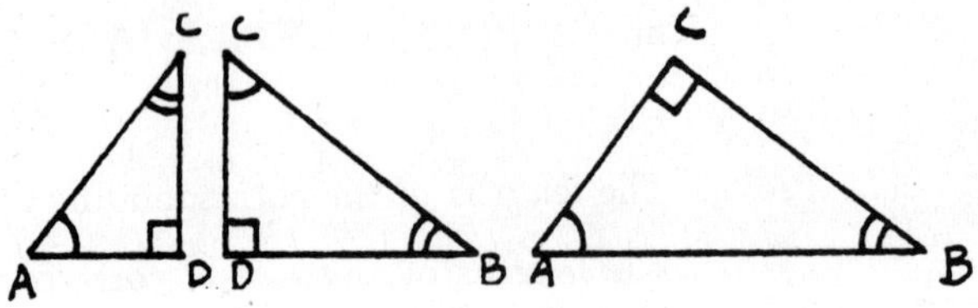

By hypothesis, $\overline{CD}$ is the altitude to $\overline{AB}$. $\overline{CD}$ is then $\perp$ to $\overline{AB}$ which means $\angle CDA$ is a rt. $\angle$. By hypothesis, $\angle ACB$ is a rt. $\angle$ in $\triangle ABC$. Therefore, $\angle CDA \doteq \angle ACB$. $\angle A$ is common to both $\triangle ADC$ and $\triangle ACB$. By AA, $\triangle ADC \sim \triangle ACB$. Since $\angle CDB$ is also a right $\angle$, $\angle CDB \doteq \angle ACD$. In $\triangle CDB$ and $\triangle ACB$, $\angle B$ is the common angle. By AA, $\triangle CDB \sim \triangle ACB$. For $\triangle CDA$ and $\triangle CDB$, both have a right $\angle$. $\angle ACD \doteq \angle CBD$ because "If 2 angles of one $\triangle$ ($\triangle ADC$) are congruent to 2 angles of another $\triangle$ ($\triangle ACB$), then the 3rd angles are also $\doteq$. Thus by AA, $\triangle ADC \sim \triangle CBD$. Hence the two right $\triangle$s formed by the altitude to the hypotenuse are similar to each other and each is similar to the original right $\triangle$.

SECTION 5.3: The Pythagorean Theorem

1. $\triangle RST \sim \triangle RVS \sim \triangle SVT$

5. a. $(DF)^2 = (DE)^2 + (EF)^2$

$(DF)^2 = 8^2 + 6^2$

$(DF)^2 = 64 + 36$

$(DF)^2 = 100$

$DF = 10$

Or (6, 8, 10) is a multiple of the Pythagorean Triple, (3,4,5); therefore, DF = 10.

b. $(DF)^2 = (DE)^2 + (EF)^2$

$(DF)^2 = 5^2 + 3^2$

$(DF)^2 = 25 + 9$

$(DF)^2 = 34$

$DF = \sqrt{34} \approx 5.83$

9. Let c be the longest side.

a. $5^2 = 3^2 + 4^2$; Right $\triangle$

b. $6^2 < 4^2 + 5^2$; Acute $\triangle$

c. $(\sqrt{7})^2 = 2^2 + (\sqrt{3})^2$; Right $\triangle$

d. No $\triangle$

13. Let x = the length of rope needed.

$x^2 = 6^2 + 12^2$

$x^2 = 36 + 144$

$x^2 = 180$

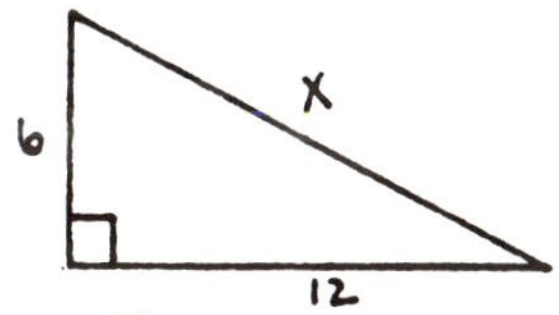

$x = \sqrt{180} = \sqrt{36 \cdot 5} = 6\sqrt{5} \approx 13.4$ meters

17. $(2x)^2 = (x + 3)^2 + (x + 1)^2$

$4x^2 = x^2 + 6x + 9 + x^2 + 2x + 1$

$4x^2 = 2x^2 + 8x + 10$

$2x^2 - 8x - 10 = 0$

$x^2 - 4x - 5 = 0$

$(x - 5)(x + 1) = 0$

$x = 5$ or $x = -1$; reject $x = -1$. The base is 8; the altitude is 6; the diagonal is 10.

21. Let the length of the hypotenuse be x

$x^2 = (6\sqrt{2})^2 + (6\sqrt{2})^2$

$x^2 = 72 + 72$

$x^2 = 144$

$x = 12$

25. A triangle whose lengths are 10, 24, and 26 is a right triangle. The altitude to the 26 inch side is the altitude to the hypotenuse. Let the lengths of the 2 segments of the hypotenuse be x and 26 − x. Let the length of the altitude to the hypotenuse be H. Using the fact that the length of each leg of a right triangle is the geometric mean for the length of the hypotenuse and the length of the segment on the hypotenuse adjacent to that leg, we have

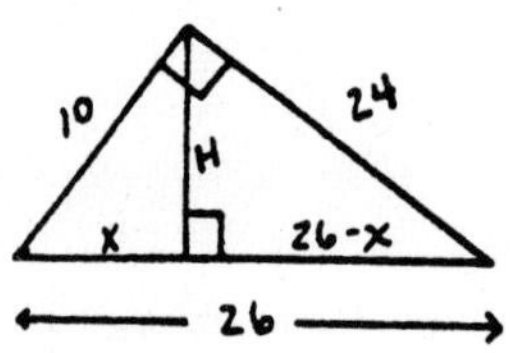

$$\frac{x}{10} = \frac{10}{26}$$

$$26x = 100$$

$$x = \frac{100}{26} = \frac{50}{13} = 3\frac{11}{13}.$$

If $x = 3\frac{11}{13}$, then $26 - x = 22\frac{2}{13}$.

Now, if the length of the altitude to the hypotenuse of a right triangle is the geometric mean of the lengths of the segments of the hypotenuse,

we have

$$\frac{\frac{50}{13}}{H} = \frac{H}{\frac{288}{13}}$$

$$H^2 = \frac{50}{13} \cdot \frac{288}{13}$$

$$H^2 = \frac{14400}{169}$$

$$H = \frac{120}{13} \text{ or } 9\frac{3}{13}.$$

25. Alternate solution:
The altitude to the 26 inch side separates it into two parts whose lengths are x and 26 − x. The length of the altitude is H. Using the Pythagorean Theorem twice, we have $x^2 + H^2 = 10^2$ and $(26 - x)^2 + H^2 = 24^2$.

Subtracting the first equation from the second gives

$$\begin{array}{r} 676 - 52x + x^2 + H = 576 \\ \underline{x^2 + H^2 = 100} \\ 676 - 52x \qquad = 476 \end{array}$$

$$-52x = -200$$

$$x = \frac{200}{52} = \frac{50}{13} = 3\ \frac{11}{13}$$

To find H, substitute $\frac{50}{13}$ for x in

$$x^2 + H^2 = 10^2$$

$$\left(\frac{50}{13}\right)^2 + H^2 = 100$$

$$\frac{2500}{169} + H^2 = 100$$

$$H^2 = 100 - \frac{2500}{169} \text{ or } \left(\frac{16900}{169} - \frac{2500}{169}\right)$$

$$H^2 = \frac{14400}{169}$$

$$H = \frac{120}{13} \text{ or } 9\frac{3}{13}$$

29. $a = p^2 - q^2$

$\therefore\ a^2 = (p^2 - q^2)^2 = p^4 - 2p^2q^2 + q^4$

$b = 2pq$

$\therefore\ b^2 = (2pq)^2 = 4p^2q^2$

$c = p^2 + q^2$

$c^2 = (p^2 + q^2)^2 = p^4 + 2p^2q^2 + q^4$

Now $a^2 + b^2 = (p^4 - 2p^2q^2 + q^4) + 4p^2q^2$

$= p^4 + 2p^2q^2 + q^4$

$= c^2$

$\therefore\ c^2 = a^2 + b^2$

33. $m\angle ACF = 60$

SECTION 5.4: Special Right Triangles

1. $YZ = 8$ and $XY = 8\sqrt{2} \approx 11.31$

5. $DF = 5\sqrt{3} \approx 8.66$ and $FE = 10$

9. In right $\triangle HLK$ if $m\angle HKL = 30$ and $LK = 6\sqrt{3}$, then $HL = 6$ and $HK = 12$.
MK = 6 since the diagonals of a rectangle bisect each other.

13. $RS = 6$ and $RT = 6\sqrt{3} \approx 10.39$

17. From vertex Z, draw an altitude to $\overline{XY}$; call the altitude $\overline{ZW}$. In the 30-60-90 $\triangle$, $WX = 6$ and $ZW = 6\sqrt{3}$. In the 45-45-90 $\triangle$, $\overline{YW} = 6\sqrt{3}$. $XY = YW + WX = 6\sqrt{3} + 6 \approx 16.39$.

21. 60°; $200^2 + x^2 = 400^2$

$x^2 = 400^2 - 200^2$

$x^2 = 120{,}000$

$x \approx 346$

The jogger travels $(200 + 346) - 400 = 146$ feet further.

25. Since $\triangle MNQ$ is equiangular and $\overrightarrow{NR}$ bisects $\angle MNQ$ and $\overrightarrow{QR}$ bisects $\angle MQN$, $m\angle RQN = 30 = m\angle RNQ$. From R, draw an altitude to $\overline{NQ}$. Name the altitude $\overline{RP}$.
$NR = RQ = 6$. In 30-60-90 $\triangle RPQ$, $RP = 3$ and $PQ = 3\sqrt{3}$. NQ therefore equals $6\sqrt{3} \approx 10.39$.

29. Draw in altitude $\overline{CD}$. In right $\triangle CDB$, if $BC = 12$, then $CD = 6$ and $DB = 6\sqrt{3}$.
In right $\triangle ACD$, if $CD = 6$, then $AD = 6$ and $AC = 6\sqrt{2}$. $AB = 6 + 6\sqrt{3} \approx 16.39$.

SECTION 5.5: Segments Divided Proportionally

1. Let $5x$ = amount of ingredient A;
$4x$ = amount of ingredient B;
$6x$ = amount of ingredient C.

$5x + 4x + 6x = 90$

$15x = 90$

$x = 6$

30 ounces of ingredient A;
24 ounces of ingredient B;
36 ounces of ingredient C.

5 . Let EF = x, FG = y, and GH = z

AD = 5 + 4 + 3 = 12 so that

$\frac{AB}{AD} = \frac{EF}{EH}$ $\frac{BC}{AD} = \frac{FG}{EH}$ $\frac{CD}{AD} = \frac{GH}{EH}$

$\frac{5}{12} = \frac{x}{10}$ $\frac{4}{12} = \frac{y}{10}$ $\frac{3}{12} = \frac{z}{10}$

$12x = 50$ $12y = 40$ $12z = 30$

$x = \frac{50}{12} = 4\frac{1}{6}$ $y = \frac{40}{12} = 3\frac{1}{3}$ $z = \frac{30}{12} = 2\frac{1}{2}$

$EF = 4\frac{1}{6}$; $FG = 3\frac{1}{3}$; $GH = 2\frac{1}{2}$.

9 . Let EC = x.

$\frac{5}{12} = \frac{7}{x}$

$5x = 84$

$x = \frac{84}{5}$ or $16\frac{4}{5}$; $EC = 16\frac{4}{5}$

13. a. No

b. Yes

17. Let NP = x = MQ.

$\frac{x}{12} = \frac{8}{x}$

$x^2 = 96$

$x = \sqrt{96} = \sqrt{16 \cdot 6} = 4\sqrt{6}$;

$NP = 4\sqrt{6} \approx 9.80$.

21. If RS = 6 and RT = 12, then $\triangle$RST is a 30-60-90 $\triangle$ and ST = $6\sqrt{3}$. Let SV = x. Then VT = $6\sqrt{3} - x$. $\frac{6}{12} = \frac{x}{6\sqrt{3} - x}$

$36\sqrt{3} - 6x = 12x$

$36\sqrt{3} = 18x$

$x = 2\sqrt{3}$. SV = $2\sqrt{3} \approx 3.46$;

VT = $6\sqrt{3} - 2\sqrt{3} = 4\sqrt{3} \approx 6.93$.

25. 1. Given
 2. Means-Extremes Property
 3. Addition Property of Equality
 4. Distributive Property
 5. Means-Extremes Property

29. Given: $\triangle$XYZ; $\overrightarrow{YW}$ bisects $\angle$XYZ
 $\overline{WX} \cong \overline{WZ}$

Prove: $\triangle$XYZ is isosceles

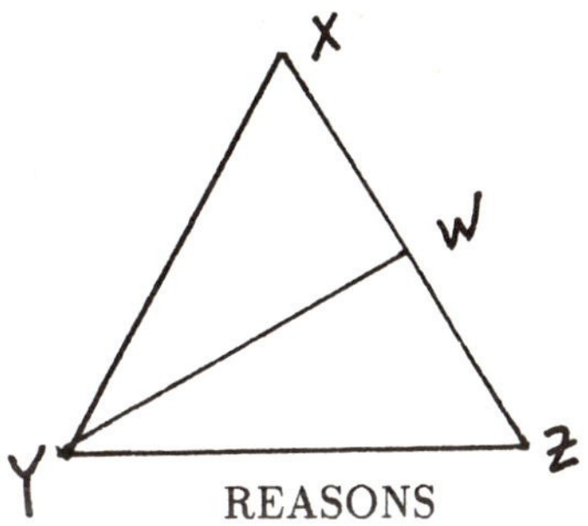

STATEMENTS	REASONS
1. $\triangle$XYZ; $\overrightarrow{YW}$ bisects $\angle$XYZ	1. Given
2. $\frac{YX}{YZ} = \frac{WX}{WZ}$	2. If a ray bisects one $\angle$ of a $\triangle$, then it divides the opposite side into segments that are proportional to the two sides which form that angle.
3. $\overline{WX} \cong \overline{WZ}$	3. Given
4. WX = WZ	4. If 2 segments are $\cong$, then their measures are equal.
5. $\frac{YX}{YZ} = \frac{WX}{WX} = 1$	5. Substitution
6. YX = YZ	6. Means-Extremes Property
7. $\overline{YX} \cong \overline{YZ}$	7. If 2 segments are equal in measure, they are $\cong$.
8. $\triangle$XYZ is isosceles	8. If 2 sides of a $\triangle$ are $\cong$, then the $\triangle$ is isosceles.

1. False
2. True
3. False
4. True
5. True
6. False
7. True
8. a. $x^2 = 18$

$x = \pm\sqrt{18}$

$= \pm\sqrt{9 \cdot 2} = \pm\ 3\sqrt{2} \approx \pm 4.24$

b. $7(x - 5) = 3(2x - 3)$

$7x - 35 = 6x - 9$

$x = 26$

c. $6(x + 2) = 2(x + 4)$

$6x + 12 = 2x + 8$

$4x = -4$

$x = -1$

d. $7(x + 3) = 5(x + 5)$

$7x + 21 = 5x + 25$

$2x = 4$

$x = 2$

e. $(x - 2)(x - 1) = (x - 5)(2x + 1)$

$x^2 - 3x + 2 = 2x^2 - 9x - 5$

$x^2 - 6x - 7 = 0$

$(x - 7)(x + 1) = 0$

$x = 7$ or $x = -1$

f. $5x(x + 5) = 9(4x + 4)$

$5x^2 + 25x = 36x + 36$

$5x^2 - 11x - 36 = 0$

$(5x + 9)(x - 4) = 0$

$x = \frac{-9}{5}$ or $x = 4$

g. $(x - 1)(3x - 2) = 10(x + 2)$

$3x^2 - 5x + 2 = 10x + 20$

$3x^2 - 15x - 18 = 0$

$x^2 - 5x - 6 = 0$

$(x - 6)(x + 1) = 0$

$x = 6$ or $x = -1$

h. $(x + 7)(x - 2) = 2(x + 2)$

$x^2 + 5x - 14 = 2x + 4$

$x^2 + 3x - 18 = 0$

$(x + 6)(x - 3) = 0$

$x = -6$ or $x = 3$

9. Let x = cost of the six containers.

$\frac{4}{2.52} = \frac{6}{x}$

$4x = 15.12$

$x = 3.78$

The six containers cost $3.78.

10. Let x = the number of packages you can buy for $2.25.

$\frac{2}{.69} = \frac{x}{2.25}$

$.69x = 4.50$

$x = \frac{450}{69} = 6\frac{12}{23}$

With $2.25, you can buy 6 packages of M & M's.

11. Let x = cost of the rug that is 12 square meters

$\frac{20}{132} = \frac{12}{x}$

$20x = 1584$

$x = 79.20$

The 12 square meters rug will cost $79.20.

12. Let the measure of the sides of the quadrilateral be 2x, 3x, 5x and 7x.

$2x + 3x + 5x + 7x = 68$

$17x = 68$

$x = 4$

The length of the sides are 8, 12, 20 and 28.

13. Let the width of the similar rectangle be x.

$\frac{18}{12} = \frac{27}{x}$

$\frac{3}{2} = \frac{27}{x}$

$3x = 54$

$x = 18$

The width of the similar rectangle is 18.

14. Let x and y be the lengths of the other two sides.

$\frac{6}{15} = \frac{8}{x} = \frac{9}{y}$

$\frac{2}{5} = \frac{8}{x}$ and $\frac{2}{5} = \frac{9}{y}$

$2x = 40$ $\quad$ $2y = 45$

$x = 20$ $\quad$ $y = \frac{45}{2} = 22\frac{1}{2}$

The other two sides have lengths 20 and $22\frac{1}{2}$.

15. Let the measure of the angle be x; the measure of the supplement would be 180 − x; the measure of the complement would be 90 − x.

$\frac{180 - x}{90 - x} = \frac{5}{2}$

$2(180 - x) = 5(90 - x)$

$360 - 2x = 450 - 5x$

$3x = 90$

$x = 30$

The measure of the supplement is 150.

16. Given: ABCD is a parallelogram;

$\overline{DB}$ intersects $\overline{AE}$ at pt. F

Prove: $\frac{AF}{EF} = \frac{AB}{DE}$

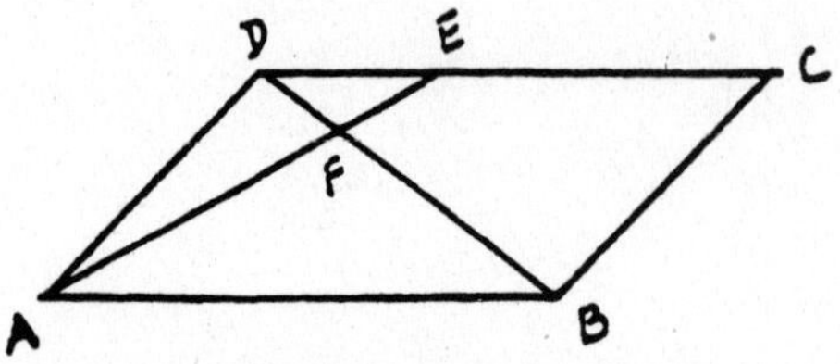

STATEMENTS	REASONS
1. ABCD is a parallelogram; $\overline{DB}$ intersects $\overline{AE}$ at pt. F	1. Given
2. $\overline{DC} \parallel \overline{AB}$	2. Opp. sides of a parallelogram are $\parallel$.
3. $\angle CDB \doteq \angle ABD$	3. If 2 $\parallel$ lines are cut by a trans., then the alt. int. $\angle$s $\doteq$.
4. $\angle DEF \doteq \angle BAF$	4. Same as (3).
5. $\triangle DFE \sim \triangle BFA$	5. AA
6. $\frac{AF}{EF} = \frac{AB}{DE}$	6. Corresp. sides of $\sim$ $\triangle$s are proportional.

17. Given: $\angle 1 \doteq \angle 2$

Prove: $\frac{AB}{AC} = \frac{BE}{CD}$

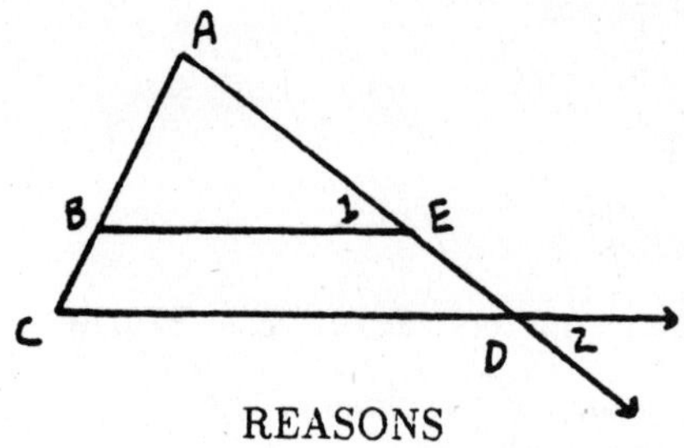

STATEMENTS	REASONS
1. $\angle ADC = \angle 2$	1. If 2 liens intersect, then the vertical formed are $\doteq$.
2. $\angle 1 \doteq \angle 2$	2. Given
3. $\angle ADC \doteq \angle 1$	3. Transitive Prop. for $\doteq$.
4. $\angle A \doteq \angle A$	4. Identity
5. $\triangle BAE \sim \triangle CAD$	5. AA
6. $\frac{AB}{AC} = \frac{BE}{CD}$	6. Corresponding sides of $\sim$ $\triangle$s are proportional.

18. Since the $\triangle$s are $\sim$, $m\angle A = m\angle D$.

$50 = 2x + 40$

$10 = 2x$

$x = 5$; $m\angle D = 2(5) + 40 = 50$;

$m\angle E = 33$;

$m\angle F = 180 - 33 - 50 = 97$

19. With $\angle B \doteq \angle F$ and

$\angle C \doteq \angle E$,

$\triangle ABC \sim \triangle DFE$.

It follows that

$\frac{AB}{DF} = \frac{AC}{DE} = \frac{BC}{FE}$

Substituting in gives

$\frac{AB}{2} = \frac{9}{3}$ or $\frac{3}{1}$ and $\frac{9}{3}$ or $\frac{3}{1} = \frac{BC}{4}$

AB = 6 and BC = 12.

20. $\frac{BD}{AD} = \frac{BE}{EC}$; let AD = x

$\frac{6}{x} = \frac{8}{4}$ or $\frac{2}{1}$

2x = 6

x = 3; AD = 3.

21. $\frac{BD}{BA} = \frac{DE}{AC}$; let AC = x; BA = 12

$\frac{8}{12} = \frac{2}{3} = \frac{3}{x}$

2x = 9

$x = 4\frac{1}{2}$; $AC = 4\frac{1}{2}$.

22. $\frac{BD}{BA} = \frac{BE}{BC}$; let BC = x; BD = 8

$\frac{8}{10}$ or $\frac{4}{5} = \frac{5}{x}$

4x = 25

$x = 6\frac{1}{4}$; $BC = 6\frac{1}{4}$.

23. Since $\overrightarrow{GJ}$ bisects $\angle FGH$, we can write the proportion

$\frac{FJ}{FG} = \frac{JH}{GH}$; let JH = x.

$\frac{7}{10} = \frac{x}{8}$

10x = 56

$x = \frac{56}{10} = 5\frac{3}{5}$; $JH = 5\frac{3}{5}$.

24. Since $\overrightarrow{GJ}$ bisects $\angle FGH$, and

GF:GH = 1:2 we can write the proportion

$\frac{GF}{GH} = \frac{FJ}{JH}$. Let JH = x

$\frac{1}{2} = \frac{5}{x}$

x = 10; JH = 10

25. Since $\overrightarrow{GJ}$ bisects $\angle FGH$, we can write the proportion

$\frac{FG}{GH} = \frac{FJ}{JH}$; let FJ = x and JH = 15 − x

$\frac{8}{12} = \frac{x}{15 - x}$

$\frac{2}{3} = \frac{x}{15 - x}$

2(15 − x) = 3x

30 − 2x = 3x

30 = 5x

x = 6; FJ = 6.

26. Let MK = x, then $\frac{MK}{HJ} = \frac{EM}{FH}$.

$\frac{x}{5} = \frac{6}{10}$ or $\frac{3}{5}$

x = 3; MK = 3

Let EO = y and OM = 6 − y.

Then $\frac{y}{2} = \frac{6 - y}{8}$

8y = 2(6 − y)

8y = 12 − 2y

10y = 12

$y = \frac{12}{10} = 1\frac{1}{5}$

$EO = 1\frac{1}{5}$; EK = 9

27. If a line bisects one side of a triangle and is parallel to a second side, then it bisects the third side.

Given: $\overleftrightarrow{DE}$ bisects $\overline{AC}$

$\overleftrightarrow{DE} \parallel \overline{BC}$

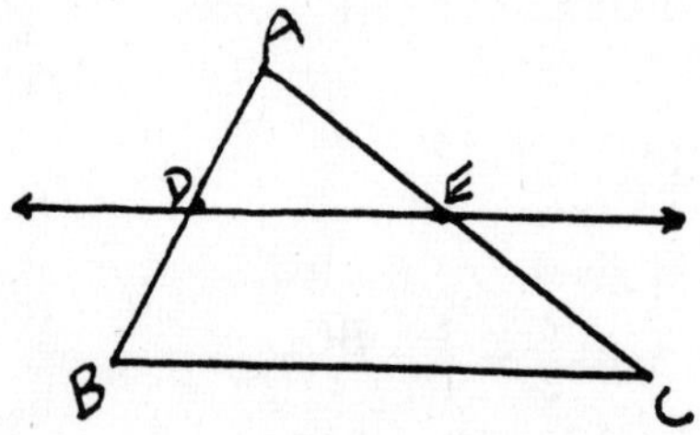

Prove: $\overleftrightarrow{DE}$ bisects $\overline{AB}$

STATEMENTS	REASONS
1. $\overleftrightarrow{DE}$ bisects $\overline{AC}$	1. Given
2. AE = EC	2. Bisecting a segment forms 2 segments of equal measure.
3. $\overleftrightarrow{DE} \parallel \overline{BC}$	3. Given
4. $\frac{AD}{DB} = \frac{AE}{EC}$	4. If a line is parallel to one side side of a triangle and intersects the other sides, then it divides these sides proportionally.
5. AD · EC = DB · AE	5. Means-Extremes Prop.
6. AD = DB	6. Division Property of Eq.
7. $\overleftrightarrow{DE}$ bisects $\overline{AB}$	7. If a segment has been divided into 2 segments of equal measure, the segment has been bisected.

28. The diagonals of a trapezoid divide themselves proportionally.

Given: ABCD is a trapezoid with $\overline{BC} \parallel \overline{AD}$ and diagonals $\overline{BD}$ and $\overline{AC}$

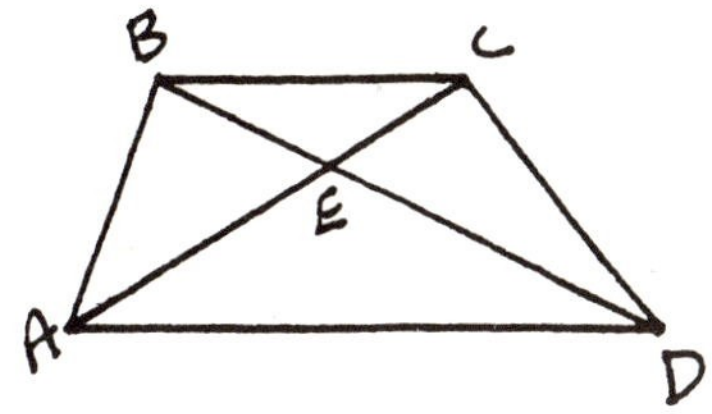

Prove: $\frac{BE}{ED} = \frac{EC}{AE}$

STATEMENTS	REASONS
1. ABCD is a trapezoid with $\overline{BC} \parallel \overline{AD}$ and diagonals $\overline{BD}$ and $\overline{AC}$.	1. Given
2. $\angle CBE \doteq \angle ADE$ and $\angle BCE \doteq \angle DAE$	2. If 2 ∥ lines are cut by a trans., then the alt. int. ∠s are ≐.
3. $\triangle BCE \sim \triangle DAE$	3. AA
4. $\frac{BE}{ED} = \frac{EC}{AE}$	4. Corresponding sides of ∼ △s are proportional.

29. a. $\frac{BD}{AD} = \frac{AD}{DC}$

Let DC = x

$\frac{3}{5} = \frac{5}{x}$

$3x = 25$

$x = \frac{25}{3}$; $DC = 8\frac{1}{3}$

b. $\frac{DC}{AC} = \frac{AC}{BC}$

Let BD = x and BC = x + 4

$\frac{4}{5}$ or $\frac{2}{5} = \frac{10}{x + 4}$

$2(x + 4) = 50$

$2x + 8 = 50$

$2x = 42$

$x = 21$; $BD = 21$

c. $\frac{BD}{BA} = \frac{BA}{BC}$

Let BA = x

$\frac{2}{x} = \frac{x}{6}$

$x^2 = 12$

$x = \sqrt{12} = \sqrt{4 \cdot 3} = 2\sqrt{3}$; $BA = 2\sqrt{3} \approx 3.46$

d. $\frac{DA}{AC} = \frac{AC}{BC}$

Let DC = x and BC = x + 3

$\frac{x}{3\sqrt{2}} = \frac{3\sqrt{2}}{2}$

$x(x + 3) = 18$

$x^2 + 3x - 18 = 0$

$(x + 6)(x - 3) = 0$

$x = -6$ or $x = 3$; reject $x = -6$.

$DC = 3$

30. a. $\frac{AD}{BD} = \frac{BD}{DC}$

Let DC= x.

$\frac{9}{12}$ or $\frac{3}{4} = \frac{12}{x}$

$3x = 48$

$x = 16$; DC = 16.

b. $\frac{DC}{BC} = \frac{BC}{AC}$

Let AD = x and AC = x + 5.

$\frac{5}{15}$ or $\frac{1}{3} = \frac{15}{x + 5}$

$x + 5 = 45$

$x = 40$; AD = 40

c. $\frac{AD}{AB} = \frac{AB}{AC}$

Let AB = x; AC = 10

$\frac{2}{x} = \frac{x}{10}$

$x^2 = 20$

$x = \sqrt{20} = \sqrt{4 \cdot 5} = 2\sqrt{5}$

$AB = 2\sqrt{5} \approx 4.47$

d. $\frac{AD}{AB} = \frac{AB}{AC}$

Let AD = x and AC = x + 2

$\frac{x}{2\sqrt{6}} = \frac{2\sqrt{6}}{x + 2}$

$x(x + 2) = 24$

$x^2 + 2x - 24 = 0$

$(x + 6)(x - 4) = 0$

$x = -6$ or $x = 4$; reject $x = -6$.

AD = 4.

31. a. x = 30. Since the leg is half of the hypotenuse, the angle opposite the leg must be 30.

b. Half of the base is 10. In the right △, 1 side has length 10 and the hypotenuse has length 26. The other side has length 24 since (10, 24, 26) is a multiple of (5, 12, 13).

c. $x^2 = 12^2 + 16^2$

$x^2 = 144 + 256$

$x^2 = 400$

$x = 20$

or (12, 16, 20) is a multiple of (3, 4, 5)

d. The unknown length of the right △ is 8 using the Triple (8, 15, 17).

$x = 16$.

32. In rect. ABCD, BC = 24 and since E is a midpoint, BE = 12 and EC = 12. CD = 16 and FD = 7. There are three right triangles for which the Pythagorean Triples apply. AE = 20 using (12, 16, 20) which is a multiple of (3, 4, 5). EF = 15 using (9, 12, 15) which is also a multiple of (3, 4, 5). AF = 25 using the Triple (7, 24, 25).

33. In a square there are two 45-45-90 △s. If the length of the side of the square is 4 inches, then the length of the diagonal is $4\sqrt{2} \approx 5.66$ inches.

34. In a square there are two 45-45-90 △s. If the length of the diagonal is 6, then $6 = a\sqrt{2}$. Solve for a gives

$a = \frac{6}{\sqrt{2}} = \frac{6\sqrt{2}}{2} = 3\sqrt{2}$

Hence, the length of a side is $3\sqrt{2} \approx 4.24$.

35. Since the diagonals of a rhombus are perpendicular and bisect each other, there are 4 right triangles formed whose sides are of lengths 24 cm and 7 cm. The hypotenuse must then have a length of 25. Since the hypotenuse of a right triangle is the side of the rhombus, the side has length 25 cm.

36. The altitude to one side of an equilateral triangle divides it into two 30-60-90 $\triangle$s. The altitude is the side opposite the 60 degree angle and is equal in length to one-half the length of the hypotenuse times $\sqrt{3}$. Hence, the altitude has length $5\sqrt{3} \approx 8.66$.

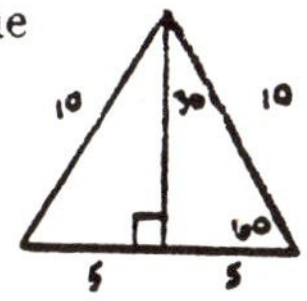

37. The altitude to one side of an equilateral triangle divides it into two 30-60-90 $\triangle$s. The altitude is the side opposite the 60 degree angle and is equal in length to one-half the length of the hypotenuse times $\sqrt{3}$. If H represents the length of the hypotenuse, then

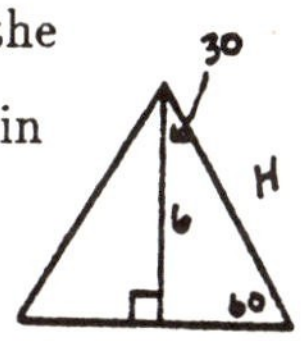

$$6 = \frac{1}{2} \cdot H \cdot \sqrt{3}$$

$$12 = H \cdot \sqrt{3}$$

$$H = \frac{12}{\sqrt{3}} \cdot \frac{\sqrt{3}}{\sqrt{3}} = \frac{12\sqrt{3}}{3} = 4\sqrt{3}$$

The length of the sides of the $\triangle$ is $4\sqrt{3} \approx 6.93$.

38. The altitude to the side of length 14 separates it into two parts; the lengths of these are given by x and 14 − x. If the length of the altitude is H, we can use the Pythagorean Theorem on the two right triangles to get

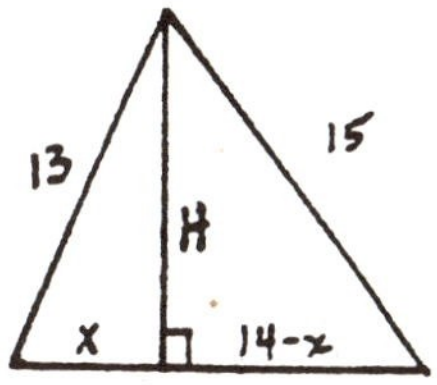

$x^2 + H^2 = 13^2$ and

$(14 - x)^2 + H^2 = 15^2$.

Subtracting the first equation from the second, we have

$$\begin{array}{rl} 196 - 29x + x^2 + H^2 &= 225 \\ x^2 + H^2 &= 169 \\ \hline 196 - 28x \qquad &= 56 \\ -28x &= -140 \\ x &= 5 \end{array}$$

Now we use x to find H.

$x^2 + H^2 = 13^2$ becomes

$$5^2 + H^2 = 169$$

$$H^2 = 144$$

$$H = 12.$$

The length of the altitude is 12 cm.

39. a. Let the length of the hypotenuse common to both $\triangle$s be H.

$\frac{1}{2} \cdot H \cdot \sqrt{3} = 9\sqrt{3}$

$H \cdot \sqrt{3} = 18\sqrt{3}$

$H = 18; y = \frac{1}{2} \cdot 18 = 9;$

$x = 9\sqrt{2} \approx 12.73.$

b. y = 6 using the Triple (6, 8, 10). Since the length of the altitude to the hypotenuse is 6, 6 is the geometric mean for x and 8. That is,

$\frac{x}{6} = \frac{6}{8}$

$8x = 36$

$x = \frac{36}{8} = \frac{9}{2} = 4\frac{1}{2}.$

c. 6 is the geometric mean for y and x. But since x = y + 9, we have

$\frac{y}{6} = \frac{6}{y+9}$

$y(y + 9) = 36$

$y^2 + 9y - 36 = 0$

$(y + 12)(y - 3) = 0$

$y = -12$ or $y = 3$

Reject $y = -12$ $\therefore$ $y = 3$ and $x = 12$.

d. $4^2 + x^2 = (6\sqrt{2})^2$

$16 + x^2 = 72$

$x^2 = 56$

$x = \sqrt{56} = \sqrt{4 \cdot 14} = 2\sqrt{14} \approx 7.48; y = 13.$

40. x = 5 using the Triple (3, 4, 5).
y = 16 using the Triple (12, 16, 20).
The ships are 11 km apart.

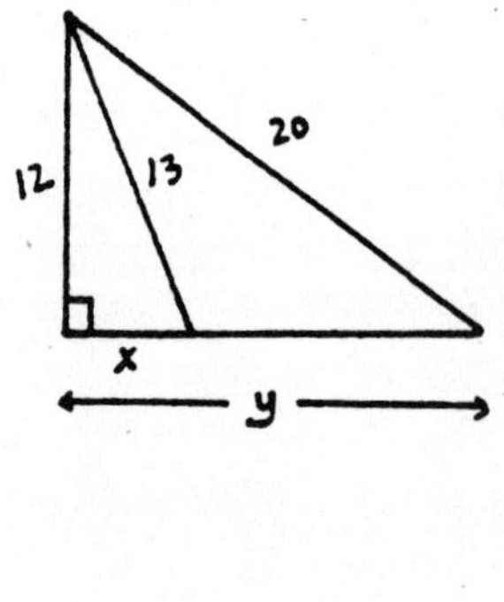

41. a. $14^2 < 12^2 + 13^2$ $\therefore$ acute $\triangle$

b. $11 + 5 \not> 18$ $\therefore$ no $\triangle$

c. $18^2 > 9^2 + 15^2$ $\therefore$ obtuse $\triangle$

d. $10^2 = 6^2 + 8^2$ $\therefore$ right $\triangle$

e. $8 + 7 \not> 16$ $\therefore$ no $\triangle$

f. $8^2 < 7^2 + 6^2$ $\therefore$ acute $\triangle$

g. $13^2 > 8^2 + 9^2$ $\therefore$ obtuse $\triangle$

h. $4^2 > 2^2 + 3^2$ $\therefore$ obtuse $\triangle$

Chapter Six: Circles

SECTION 6.1: Circles and Related Segments and Angles

1. 29°
5. 56.6°
9. Since the $m\overset{\frown}{AB} : m\overset{\frown}{BC} : m\overset{\frown}{CA} = 2{:}3{:}4$, we can write the equation $2x + 3x + 4x = 360$. Solving for x we get $9x = 360$ or $x = 40$.

   a. 80 b. 120
   c. 160 d. 80
   e. 120 f. 160
   g. 10 h. 50
   i. 30

13. a. 12 b. $6\sqrt{2}$
17. Draw in $\overline{AO}$, $\overline{BO}$, $\overline{AQ}$, and $\overline{BQ}$. $\triangle OAQ \doteq \triangle OBQ$ by SSS. If $\angle AOB \doteq \angle BOQ$ by CPCTC, then $\triangle AOR \doteq \triangle BOR$ by SAS. $\overline{AR} \doteq \overline{BR}$ and $\angle QRA \doteq \angle QRB$ by CPCTC. Hence, $\overline{OQ}$ is the perpendicular bisector of $\overline{AB}$. In right triangle OAR, AR = 3 and OA = 4. Using the Pythagorean Theorem, we have $3^2 + (OR)^2 = 4^2 \rightarrow 9 + (OR)^2 = 16 \rightarrow (OR)^2 = 7 \rightarrow OR = \sqrt{7}$. In rt. $\triangle QAR$, AQ = 6 and AR = 3. Using the Pythagorean Theorem again, we have $3^2 + (RQ)^2 = 6^2 \rightarrow 9 + (RQ)^2 = 36 \rightarrow (RQ)^2 = 27 \rightarrow RQ = \sqrt{27} = 3\sqrt{3}$.
21. a. The measure of an arc equals the measure of its intercepted arc. Therefore, congruent arcs would have to have congruent central angles.

    b. The measure of a central angle equals the measure of its intercepted arc. Therefore, congruent central angles have congruent arcs.

    c. Draw the radii to the endpoints of the congruent chords. The two triangles formed are congruent by SSS. The central angles of each triangle are congruent by CPCTC. Therefore, the arcs corresponding to the central angles are also congruent. Hence, congruent chords will congruent arcs.

    d. Draw the four radii to the endpoints of the congruent arcs. Also draw the chords corresponding to the congruent arcs. The central angles corresponding to the congruent arcs are also congruent. Therefore, the triangles are congruent by SAS. The chords are congruent by CPCTC. Hence, congruent arcs will have congruent chords.

    e. Congruent central angles will have congruent arcs. (from b) Congruent arcs will have congruent chords. (from d) Hence, congruent central angles will have congruent chords.

    f. Congruent chords have congruent arcs. (from c) Congruent arcs have congruent central angles. (from a) Therefore, congruent chords will have congruent central angles.

25. 72°

29. 1. $\overline{MN} \parallel \overline{OP}$ in $\odot O$

2. If 2 $\parallel$ lines are cut by a transversal, then the alt. int. $\angle$s are $\doteq$.

3. If 2 $\angle$s are $\doteq$, then their measures are =.

4. The measure of an inscribed $\angle$ = $\frac{1}{2}$ the measure of its intercepted arc.

5. The measure of a central $\angle$ equals the measure of its arc.

6. Substitution

33. Prove: A radius perpendicular to a chord bisects the arc of that chord.

Given: $\overline{OC} \perp \overline{AB}$ in $\odot O$

Prove: $\overline{OC}$ bisects AB

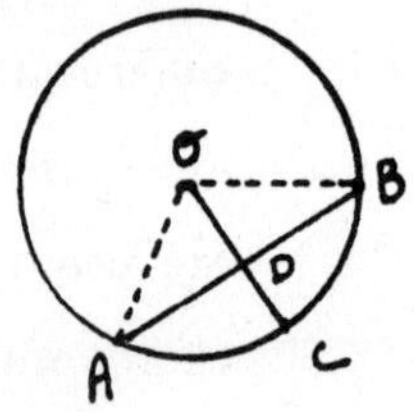

Proof: If $\overline{OC} \perp \overline{AB}$ in $\odot O$, then $\overline{OC}$ bisects $\overline{AB}$ giving $\overline{AD} \doteq \overline{DB}$. Draw in radii $\overline{OA}$ and $\overline{OB}$ which are congruent. Using $\overline{OD} \doteq \overline{OD}$, $\triangle ODA \doteq \triangle ODB$ by SSS. Therefore, $\angle AOC \doteq \angle BOC$ by CPCTC. $\overset{\frown}{AC} \doteq \overset{\frown}{CB}$ since congruent central angles will have congruent arcs. Hence, $\overline{OC}$ bisects $\overset{\frown}{AB}$.

37. If $\overset{\frown}{ST} \doteq \overset{\frown}{TV}$, $\overline{ST} \doteq \overline{TV}$ since $\doteq$ arcs in a circle have $\doteq$ chords. $\triangle STV$ is an isosceles $\triangle$ because it has two $\doteq$ sides.

SECTION 6.2: More Angle Measures in the Circle

1 . If $m\overset{\frown}{AB} = 92$, $m\overset{\frown}{DA} = 114$, and $m\overset{\frown}{BC} = 138$, then the $m\overset{\frown}{DC} = 16$.

a. $m\angle 1 = \frac{1}{2}(16) = 8$

b. $m\angle 2 = \frac{1}{2}(92) = 46$

c. $m\angle 3 = \frac{1}{2}(92 - 16) = \frac{1}{2}(76) = 38$

d. $m\angle 4 = \frac{1}{2}(92 + 16) = \frac{1}{2}(108) = 54$

e. $m\angle 5 = 180 - 54 = 126$ or

$m\angle 5 = \frac{1}{2}(114 + 138) = \frac{1}{2}(252) = 126$

5 . If $m\overset{\frown}{MP} = 112$ and $m\overset{\frown}{MN} = 60$, then $m\overset{\frown}{NP} = 52$.

If $\overline{PV}$ is a diameter in $\odot Q$, then $m\overset{\frown}{MV} = 68$.

If $m\overset{\frown}{MT} = 46$, then $m\overset{\frown}{TV} = 22$.

a. $m\angle MRP = \frac{1}{2}(112 - 68) = \frac{1}{2}(44) = 22$

b. $m\angle 1 = \frac{1}{2}(60 - 46) = \frac{1}{2}(14) = 7$

c. $m\angle 2 = \frac{1}{2}(52 - 22) = \frac{1}{2}(30) = 15$

9 . $m\angle 2 = \frac{1}{2}(m\overset{\frown}{AB} - m\overset{\frown}{DC})$

$36 = \frac{1}{2}(4x - x)$

$36 = \frac{1}{2}(3x)$

$72 = 3x$

$24 = x$

a. $m\overset{\frown}{AB} = 4(24) = 96$

b. $m\angle 1 = \frac{1}{2}(96 + 24)$

$= \frac{1}{2}(120)$

$= 60$

13. If $m\angle 2 = 124$, then $m\angle 1 = 56$ because they are supplementary $\angle$s.

$56 = \frac{1}{2}[1x + 1 + 3(x + 1)]$

$112 = 1x + 1 + 3x + 3$

$108 = 4x$

$x = 27$; $m\overset{\frown}{TV} = 27 + 1 = 28$

17. 1. $\overline{AB}$ and $\overline{AC}$ are tangents to $\odot O$ from A

2. Measure of an $\angle$ formed by a tangent and a chord = $\frac{1}{2}$ the arc measure.

3. Substitution

4. If 2 $\angle$s are = in measure, they are $\doteq$.

5. $\overline{AB} \doteq \overline{AC}$

6. $\triangle ABC$ is isosceles

21. If $m\overset{\frown}{AB} = x$, then $m\overset{\frown}{ADB} = 360 - x$. Then

$m\angle 1 = \frac{1}{2}(m\overset{\frown}{ADB} - m\overset{\frown}{AB})$

$= \frac{1}{2}(360 - x - x)$

$= \frac{1}{2}(360 - 2x)$

$= 180 - x$

25. Each arc of the circle is 72.

$m\angle 1 = \frac{1}{2}(72) = 36.$

$m\angle 2 = \frac{1}{2}(144 + 72) = \frac{1}{2}(216) = 108.$

29. Using the given information, we know that $\angle ACB$ is a right $\angle$. Let $AM = MC = x$ and $CN = NB = y$.

Using the Pythagorean Theorem we have,

$(x)^2 + (2y)^2 = (\sqrt{73})^2$ and

$(2x)^2 + (y)^2 = (2\sqrt{13})^2$

$x^2 + 4y^2 = 73$ (Multiply by -4)

$4x^2 + y^2 = 52$

$$\begin{array}{r} -4x^2 - 16y^2 = -292 \\ 4x^2 + y^2 = 52 \\ \hline -15y^2 = -240 \end{array}$$

$y^2 = 16$

$y = 4$

$x^2 + 4(4)^2 = 73$

$x^2 + 4(16) = 73$

$x^2 + 64 = 73$

$x^2 = 9$

$x = 3$

$AM = MC = 3$ and $CN = NB = 4$

Therefore $AC = 6$ and $CB = 8$, $AB = 10$ using the Pythagorean Triple (6, 8, 10).

33. If a trapezoid is inscribed in a circle, then it is an isosceles trapezoid.

Given: ABCD is an inscribed trapezoid.

Prove: ABCD is an isosceles trapezoid.

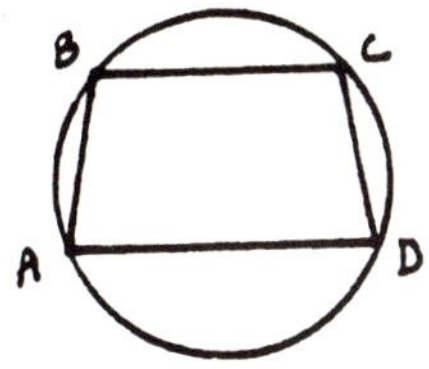

Proof: If ABCD is a trapezoid, then $\overline{BC} \parallel \overline{AD}$. If 2 $\parallel$ lines intersect a circle, then the intercepted arcs between these lines are congruent. Therefore $\overset{\frown}{AB} \cong \overset{\frown}{CD}$, which means $\overline{AB} \cong \overline{CD}$. If the nonparallel sides of a trapezoid are congruent, then trapezoid ABCD is an isosceles trapezoid.

37. Given: $\overline{AC}$ is a diameter in $\odot O$ and $\overrightarrow{CD}$ is a tangent.

Prove: $m\angle BCD = \frac{1}{2}m\overset{\frown}{BAC}$

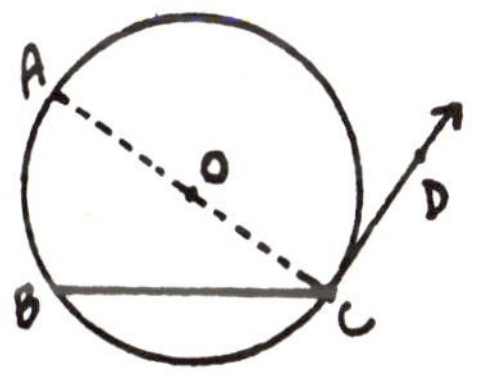

Proof: $\overline{AC}$ is a diameter in $\odot O$ and $\overline{CD}$ is a tangent.

$m\angle BCD = m\angle BCA + m\angle ACD$

(Angle Addition Postulate)

$m\angle ACD = \frac{1}{2}m\overset{\frown}{AC}$ (Case (1))

$m\angle BCA = \frac{1}{2}m\overset{\frown}{BA}$ (Inscribed $\angle$)

$m\angle BCD = \frac{1}{2}m\overset{\frown}{BA} + \frac{1}{2}m\overset{\frown}{AC} = \frac{1}{2}(m\overset{\frown}{BA} + m\overset{\frown}{AC})$

(Substitution and Distributive)

$m\overset{\frown}{BA} + m\overset{\frown}{AC} = m\overset{\frown}{BAC}$

(Arc-Addition Postulate)

$m\angle BCD = \frac{1}{2}m\overset{\frown}{BAC}$ (Substitution)

SECTION 6.3: Line and Segment Relationships in the Circle

1. $\triangle OCD$ is an equilateral triangle with $m\angle COD = 60$. Since $\overline{OE} \perp \overline{CD}$, $\overline{OF}$ bisects $\overset{\frown}{CD}$. $m\overset{\frown}{CF} = m\overset{\frown}{FD}$ and $m\angle COF = m\angle FOD = 30$. $m\overset{\frown}{CF}$ must equal 30.

5. a.

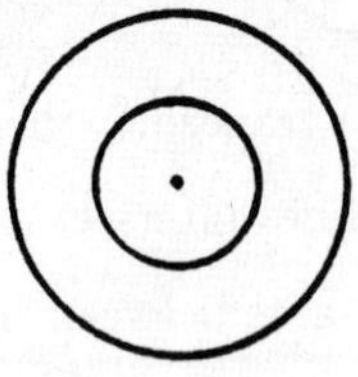

b.

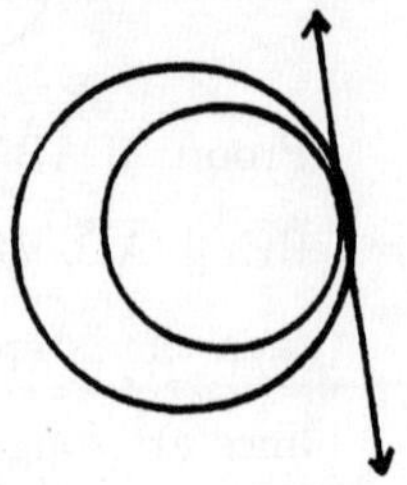

c.

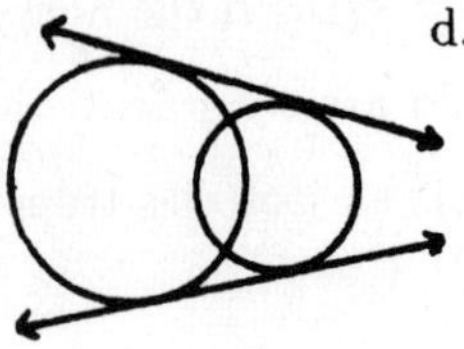

d.

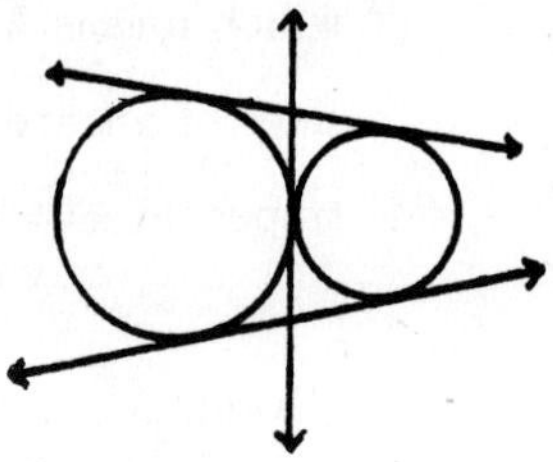

e.

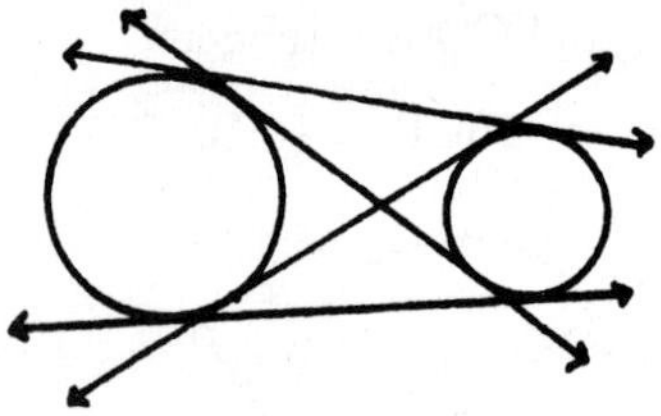

9. Let DE = x and EC = 16 − x. Then

$8 \cdot 6 = x(16 - x)$

$48 = 16x - x^2$

$x^2 - 16x + 48 = 0$

$(x - 4)(x - 12) = 0$

$x = 4$ or $x = 12$

DE = 4 and EC = 12 OR

DE = 12 and EC = 4.

13. If AB = 6 and BC = 8, then AC = 14.

Let DE = x and AD = 15 − x.

$6 \cdot 14 = (15 - x) \cdot 15$

$84 = 225 - 15x$

$-141 = -15x$

$x = 9\frac{2}{5}$; DE = $9\frac{2}{5}$

17. Let RT = x.

$8^2 = x \cdot 12$

$64 = 12x$

$x = 5\frac{1}{3}$; RT = $5\frac{1}{3}$.

21. a. None b. One c. 4

25. Let M, N, P, and Q be the pts. of tangency for $\overline{DC}$, $\overline{DA}$, $\overline{AB}$, and $\overline{BC}$, respectively. Because tangent segments from an ext. point are $\cong$, AP = AN, PB = BQ, CM = CQ, and MD = DN. ∴ AP + PB + CM + MD = AN + BQ + CQ + DN. Reordering and associating, (AP + PB) + (CM + MD) = (AN + DN) + (BQ + CQ) = AB + CD = DA + BC.

29. $m\angle D = \frac{1}{2}(200 - 160) = 20°$

33. Draw in $\overline{AO}$, $\overline{BQ}$, and $\overline{OQ}$. Since $\overline{AO} \perp \overline{AB}$ and $\overline{BQ} \perp \overline{AB}$, ABQO is a trapezoid. Now draw a segment from O perpendicular to $\overline{BQ}$. Call the point of intersection R. Now ABRO is a rectangle. If OC = OA = 4, and BQ = CQ = 9, then OQ = 13 and RQ = 5. But $\triangle ORQ$ is a right $\triangle$ and OR = 12 by the Pythagorean Triple (5, 12, 13). AB must also be 12 since ABRO is a rectangle.

37. Let x represent the angle measure of the larger gear. It is intuitively obvious that the

$$\frac{\text{number of teeth in the larger gear}}{\text{number of teeth in the smaller gear}} = \frac{\text{angle measure in smaller gear}}{\text{angle measure in larger gear}}.$$

41. Given: $\overline{TX}$ is a secant segment and $\overline{TV}$ is a tangent at V

Prove: $(TV)^2 = TW \cdot TX$

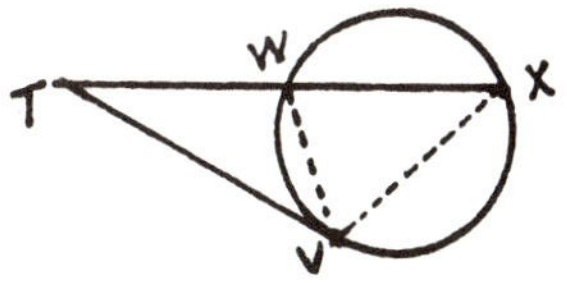

Proof: With secant $\overline{TX}$ and tangent $\overline{TV}$, draw in $\overline{WV}$ and $\overline{VX}$. $m\angle X = \frac{1}{2}m\overset{\frown}{WV}$ since $\angle X$ is an inscribed angle.

$m\angle TVW = \frac{1}{2}m\overset{\frown}{WV}$ because it is formed by a tangent and a chord. By substitution, $m\angle TVW = m\angle X$ or $\angle TVW \cong \angle X$.

$\angle T \cong \angle T$ and $\triangle TVW \sim \triangle TXV$.

It follows that $\frac{TV}{TW} = \frac{TX}{TV}$ or $(TV)^2 = TW \cdot TX$.

SECTION 6.4: Some Constructions and Inequalities for the Circle

1. $m\angle CQD < m\angle AQB$

5. $CD < AB$

9. No

13. $\overline{AB}$; $\overline{GH}$; for a circle containing unequal chords, the chord nearest the center has the greatest length and the chord at the greatest distance from the center has the least length.

17. a. $m\overset{\frown}{MN} > m\overset{\frown}{QP}$ b. $m\overset{\frown}{MPN} < m\overset{\frown}{PMQ}$

21. a. $m\angle AOB > m\angle BOC$ b. $AB > BC$

25. a. $\angle C$ b. $\overline{AC}$

29.

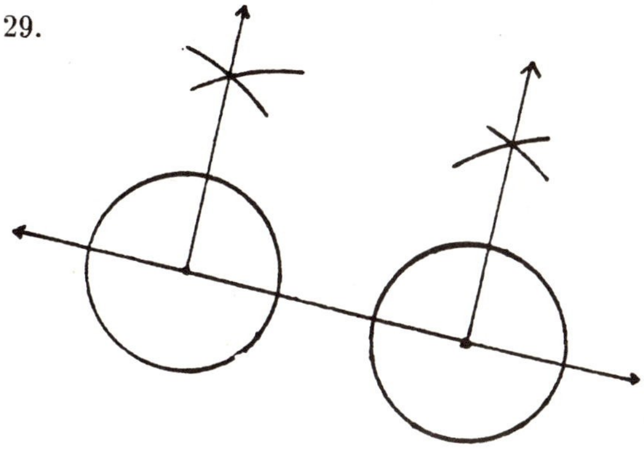

33. Draw in $\overline{OB}$ to form rt. $\triangle$ OMB. Since $MB = 12$ and $OB = 13$, $OM = 5$. Draw in $\overline{OD}$ to form rt. $\triangle$OND. Since $ND = 5$ and $OD = 13$, $ON = 12$.

$MN = ON - OM$; $MN = 12 - 5 = 7$.

SECTION 6.5: Locus of Points

1.

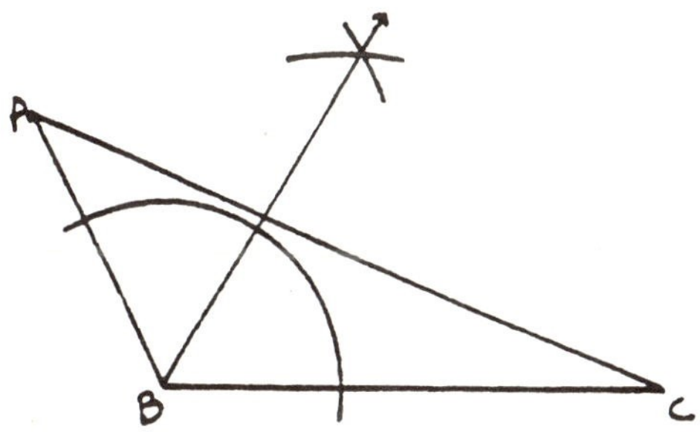

5.

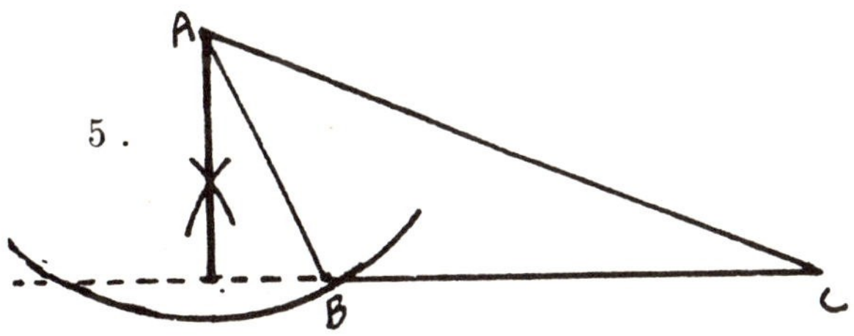

9. The locus of points at a given distance from a fixed line is two parallel lines on either side of the fixed line at the same distance from the fixed line.

13. The locus of points equidistant from the three noncollinear points D, E, and F is the circumcenter of $\triangle$ DEF.

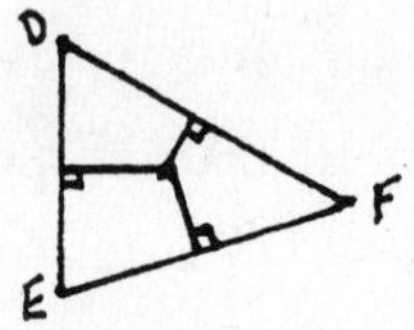

17. The locus of points equidistant from two given intersecting lines are two perpendicular lines which bisect the angles formed by the 2 intersecting lines.

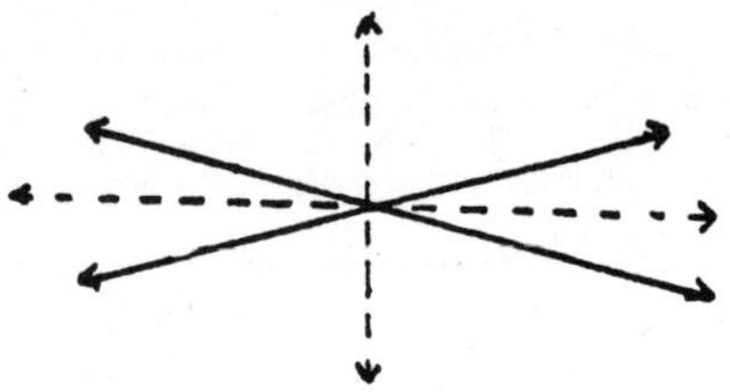

21. The locus of points at a given distance from a fixed line is a cylinder (like a tin can without a lid or base).

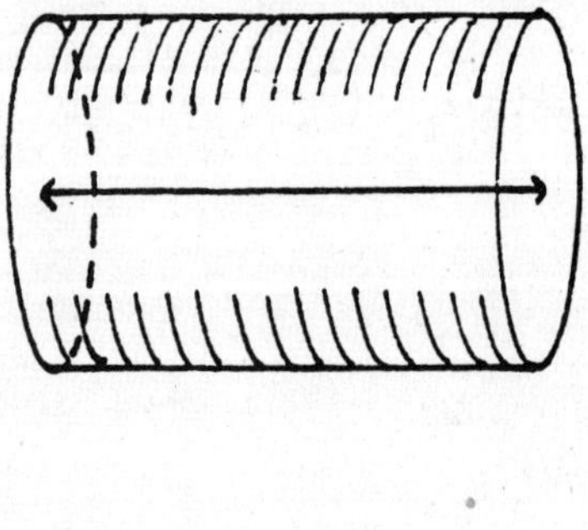

25. The locus is another sphere with the same center and a radius of length 2.5 meters.

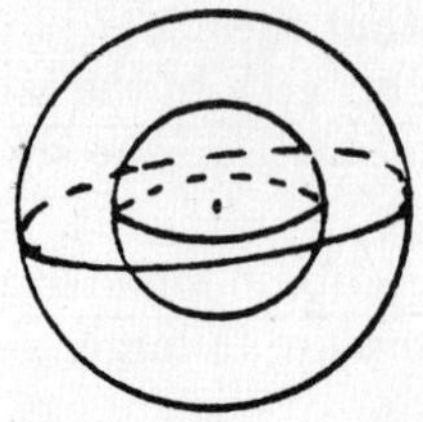

29. By the method of construction and the fact that "all radii of a circle are equal in length," we know that $\overline{AC} \doteq \overline{BC}$ and $\overline{AD} \doteq \overline{BD}$. Also $\overline{CD} \doteq \overline{CD}$ and it follows that $\triangle CAD \doteq \triangle CBD$ by SSS. Then $\angle ACD \doteq \angle BCD$ by CPCTC. Using $\overline{CE} \doteq \overline{CE}$, $\triangle ACE \doteq \triangle BCE$ by SAS. It follows that $\overline{AE} \doteq \overline{EB}$ and $\angle AEC \doteq \angle BEC$ by CPCTC. Hence, $\overline{CD}$ is the perpendicular bisector of $\overline{AB}$.

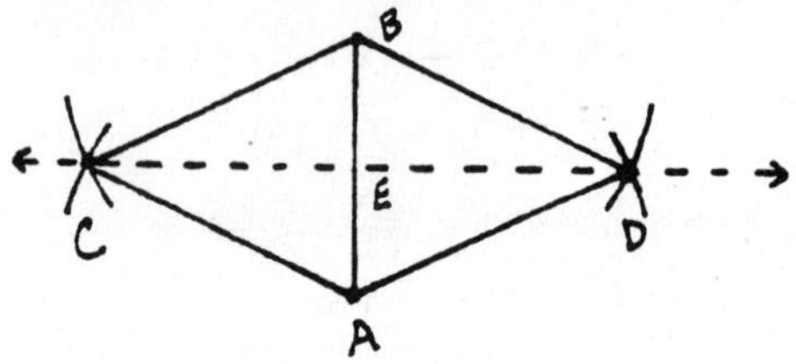

33.

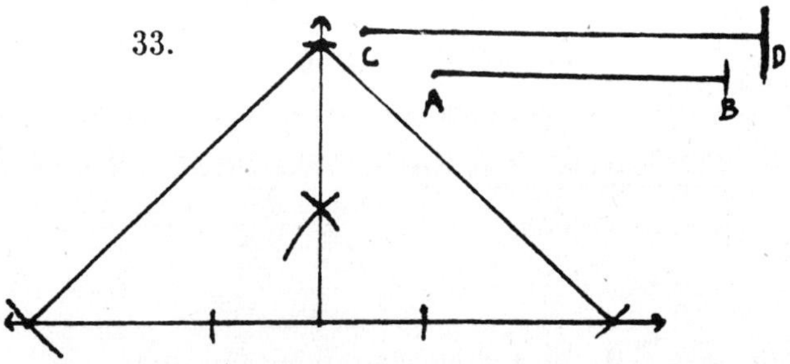

37.

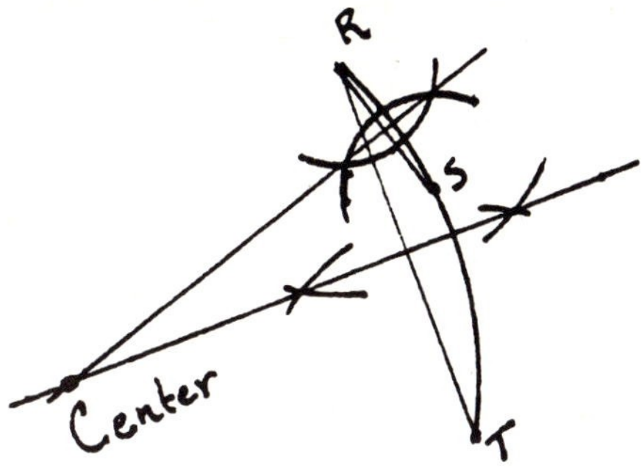

SECTION 6.6: Concurrence of Lines

1. a. Angle bisectors
   b. Perpendicular bisectors
   c. Altitudes
   d. Medians

5. Equilateral triangle

9.

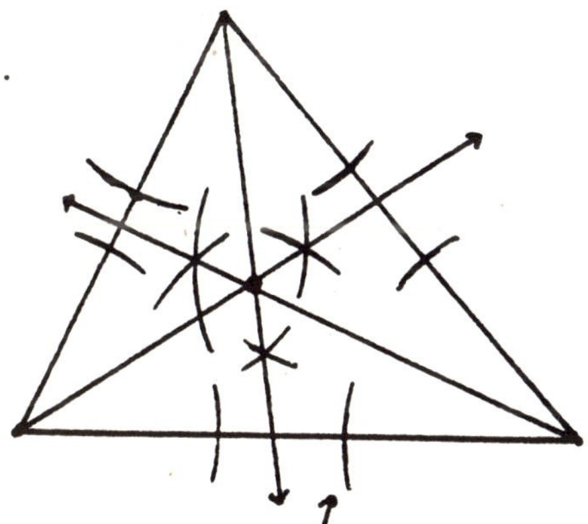

13.

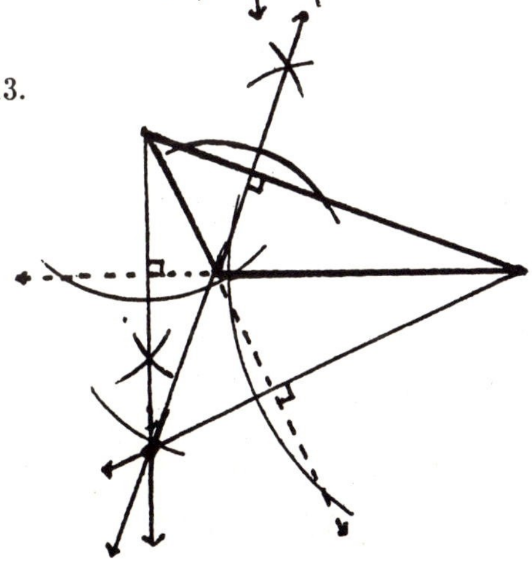

17. No

21. RZ = 15 using the Pythagorean Triple (8,15,17). RQ = $\frac{2}{3} \cdot 15 = 10$.
In right $\triangle QZS$, QZ = 5 and ZS = 8.
Using the Pythagorean Theorem, we have
$(SQ)^2 = 8^2 + 5^2$
$(SQ)^2 = 64 + 25$
$(SQ)^2 = 89$
$SQ = \sqrt{89}$.

25.

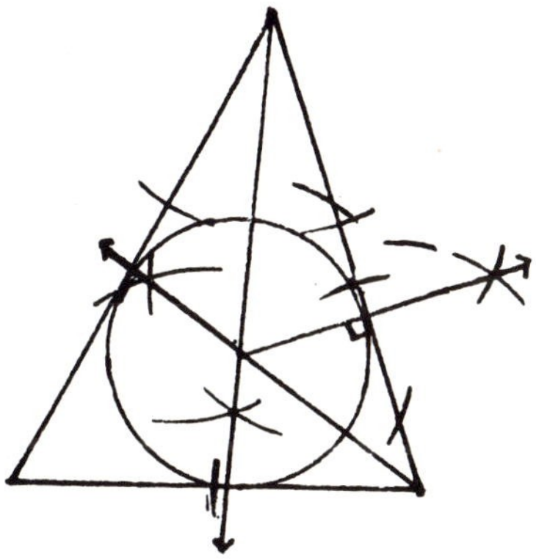

29. a. Yes b. Yes

33.

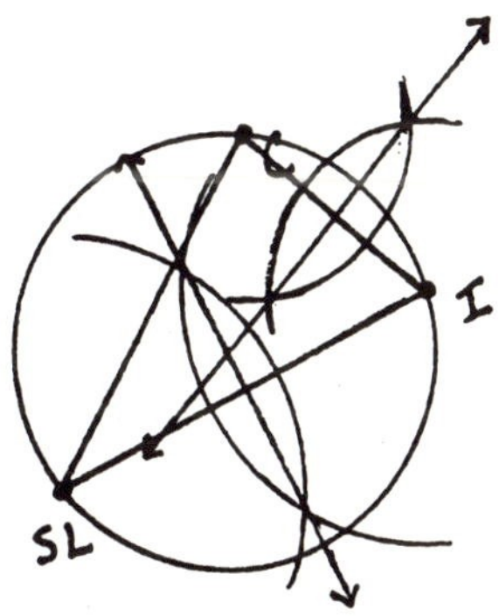

1. (9, 12, 15) is a multiple of the Pythagorean Triple (3,4,5). Therefore, the distance from the center of the circle to the chord is 9 mm.

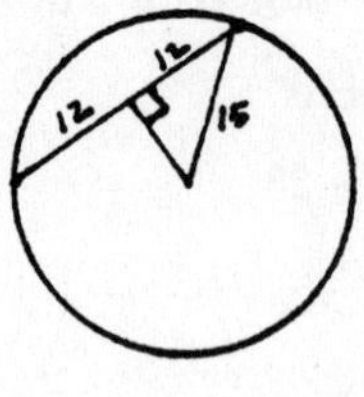

2. (8, 15, 17) is a Pythagorean Triple. Therefore, the length of half of the chord is 15 and the length of the chord is 30 cm.

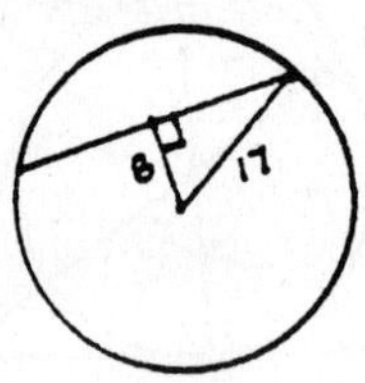

3. $r^2 = 5^2 + 4^2$

$r^2 = 25 + 16$

$r^2 = 41$

$r = \sqrt{41}$

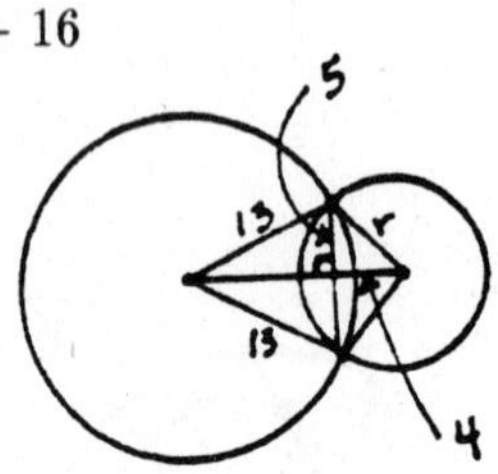

4. The radius of each circle has a length of $6\sqrt{2}$.

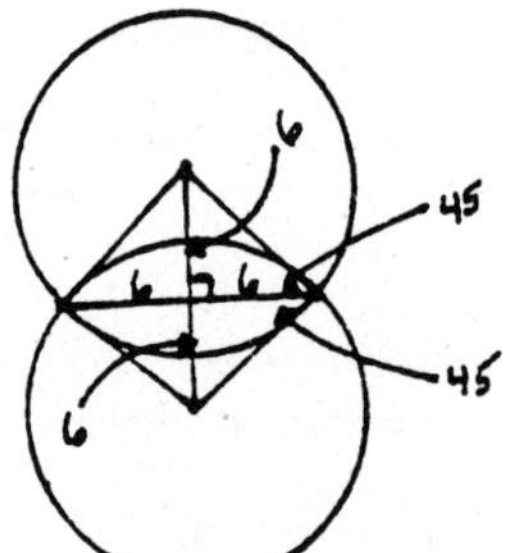

5. $m\angle B = \frac{1}{2}(m\overset{\frown}{AD} - m\overset{\frown}{AC})$

$25 = \frac{1}{2}(140 - m\overset{\frown}{AC})$

$50 = 140 - m\overset{\frown}{AC}$

$m\overset{\frown}{AC} = 90$

$m\overset{\frown}{DC} = 360 - (140 + 90) = 130$

6. $m\overset{\frown}{AC} = 360 - (190 + 120) = 50$

$m\angle B = \frac{1}{2}(m\overset{\frown}{AD} - m\overset{\frown}{AC})$

$m\angle B = \frac{1}{2}(120 - 50) = \frac{1}{2}(70) = 35$

7. If $m\angle EAD = 70$, the $m\overset{\frown}{AD} = 140$.

$m\angle B = \frac{1}{2}(m\overset{\frown}{AD} - m\overset{\frown}{AC})$

$30 = \frac{1}{2}(140 - m\overset{\frown}{AC})$

$60 = 140 - m\overset{\frown}{AC}$

$m\overset{\frown}{AC} = 80$

8. If $m\angle D = 40$, then $m\overset{\frown}{AC} = 80$.

$m\overset{\frown}{AD} = 360 - (80 + 130) = 150$

$m\angle B = \frac{1}{2}(m\overset{\frown}{AD} - m\overset{\frown}{AC})$

$m\angle B = \frac{1}{2}(150 - 80)$

$m\angle B = \frac{1}{2}(70) = 35$

9. Let $m\overset{\frown}{AC} = m\overset{\frown}{CD} = x$. Then

$m\overset{\frown}{AD} = 360 - 2x.$

$m\angle B = \frac{1}{2}(m\overset{\frown}{AD} - m\overset{\frown}{AC})$

$40 = \frac{1}{2}(360 - 2x - x)$

$80 = 360 - 3x$

$-280 = -3x$

$x = 93\frac{1}{3}$

$m\overset{\frown}{AC} = m\overset{\frown}{DC} = 93\frac{1}{3};$

$m\overset{\frown}{AD} = 173\frac{1}{3}$

10. Let $m\overset{\frown}{AC} = x$; $m\overset{\frown}{AD} = 290 - x$

$m\angle B = \frac{1}{2}(m\overset{\frown}{AD} - m\overset{\frown}{AC})$

$35 = \frac{1}{2}(290 - x - x)$

$70 = 290 - 2x$

$-220 = -2x$

$x = 110$

$m\overset{\frown}{AC} = 110$ and $m\overset{\frown}{AD} = 180$

11. If $m\angle 1 = 46$, then $m\overset{\frown}{BC} = 92$.

If $\overline{AC}$ is a diameter, $m\overset{\frown}{AB} = 88$.

$m\angle 2 = 44$; $m\angle 3 = 90$; $m\angle 4 = 46$;

$m\angle 5 = 44$.

12. If $m\angle 5 = 40$, then $m\overset{\frown}{AB} = 80$.

If $\overline{AC}$ is a diameter, $m\overset{\frown}{BC} = 100$.

$m\angle 1 = 50$; $m\angle 2 = 40$; $m\angle 3 = 90$,

$m\angle 4 = 50$.

13. (12, 16, 20) is a multiple of the Pythagorean Triple (3,4,5). Hence, half of the chord has a length of 12 and the chord has length 24.

14. The radius of the circle has length 10 using the Pythagorean Triple (6, 8, 10).

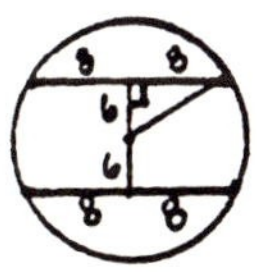

15. A

16. S

17. N

18. S

19. A

20. N

21. A

22. N

23. a. $m\angle AEB = \frac{1}{2}(m\overset{\frown}{AB} + m\overset{\frown}{CD})$

$75 = \frac{1}{2}(80 + m\overset{\frown}{CD})$

$150 = 80 + m\overset{\frown}{CD}$

$m\overset{\frown}{CD} = 70$

b. $m\angle BED = \frac{1}{2}(m\overset{\frown}{AC} + m\overset{\frown}{BD})$

$45 = \frac{1}{2}(62 + m\overset{\frown}{BD})$

$90 = 62 + m\overset{\frown}{BD}$

$m\overset{\frown}{BD} = 28$

c. $m\angle P = \frac{1}{2}(m\overset{\frown}{AB} - m\overset{\frown}{CD})$

$24 = \frac{1}{2}(88 - m\overset{\frown}{CD})$

$48 = 88 - m\overset{\frown}{CD}$

$m\overset{\frown}{CD} = 40$

$m\angle CED = \frac{1}{2}(m\overset{\frown}{AB} + m\overset{\frown}{CD})$

$m\angle CED = \frac{1}{2}(88 + 40) = \frac{1}{2}(128) = 64$

d. $m\angle CED = \frac{1}{2}(m\overset{\frown}{AB} + m\overset{\frown}{CD})$

$41 = \frac{1}{2}(m\overset{\frown}{AB} + 20)$

$82 = m\overset{\frown}{AB} + 20$

$m\overset{\frown}{AB} = 62$

$m\angle P = \frac{1}{2}(m\overset{\frown}{AB} - m\overset{\frown}{CD})$

$m\angle P = \frac{1}{2}(62 - 20) = \frac{1}{2}(42) = 21$

e. $m\angle AEB = \frac{1}{2}(m\overset{\frown}{AB} + m\overset{\frown}{CD})$ and

$m\angle P = \frac{1}{2}(m\overset{\frown}{AB} - m\overset{\frown}{CD})$

$65 = \frac{1}{2}(m\overset{\frown}{AB} + m\overset{\frown}{CD})$

$25 = \frac{1}{2}(m\overset{\frown}{AB} - m\overset{\frown}{CD})$

$$\begin{array}{rl} & 130 = m\overset{\frown}{AB} + m\overset{\frown}{CD} \\ - & \underline{50 = m\overset{\frown}{AB} - m\overset{\frown}{CD}} \\ & 180 = 2 \cdot m\overset{\frown}{AB} \end{array}$$

$mAB = 90$; $m\overset{\frown}{CD} = 40$

23. f. $m\angle CED = \frac{1}{2}(m\overset{\frown}{AB} + m\overset{\frown}{CD})$

$50 = \frac{1}{2}(m\overset{\frown}{AB} + m\overset{\frown}{CD})$

$100 = m\overset{\frown}{AB} + m\overset{\frown}{CD}$

$m\overset{\frown}{AC} + m\overset{\frown}{BD} = 360 - 100 = 260$

24. a. Let BC = x.

$6^2 = 12 \cdot x$

$36 = 12x$

$x = 3$; BC = 3

b. Let DG = x

$4 \cdot 6 = x \cdot 3$

$24 = 3x$

$x = 8$; DG = 8

c. Let CE = x

$3 \cdot x = 4 \cdot 12$

$3x = 48$

$x = 16$; CE = 16

d. Let GE = x

$10 \cdot x = 5 \cdot 8$

$10x = 40$

$x = 4$; GE = 4

e. Let BC = x and

CA = x + 5

$6^2 = x(x + 5)$

$36 = x^2 + 5x$

$0 = x^2 + 5x - 36$

$0 = (x + 9)(x - 4)$

$x = -9$ or $x = 4$;

Reject -9; BC = 4.

f. Let DG = x

and AG = 9 − x

$x(9 - x) = 4 \cdot 2$

$9x - x^2 = 8$

$0 = x^2 - 9x + 8$

$0 = (x - 8)(x - 1)$

$x = 8$ or $x = 1$; GD = 8

or GD = 1.

g. Let CD = ED = x and CE = 2x

$x(2x) = 3 \cdot 30$

$2x^2 = 90$

$x^2 = 45$

$x = \sqrt{45} = \sqrt{9 \cdot 5} = 3\sqrt{5}$; ED = $3\sqrt{5}$

h. Let CD = x and CE = x + 12

$x(x + 12) = 5 \cdot 9$

$x^2 + 12x = 45$

$x^2 + 12x - 45 = 0$

$(x + 15)(x - 3) = 0$

$x = -15$ or $x = 3$; reject $x = -15$; CD = 3

i. Let FC = x

$x^2 = 4 \cdot 12$

$x^2 = 48$

$x = \sqrt{48} = \sqrt{16 \cdot 3} = 4\sqrt{3}$; FC = $4\sqrt{3}$

j. Let CD = x and CE = x + 9

$x(x + 9) = 6^2$

$x^2 + 9x = 36$

$x^2 + 9x - 36 = 0$

$(x + 12)(x - 3) = 0$

$x = -12$ or $x = 3$; reject -12; CD = 3

25. $5x + 4 = 2x + 19$

$3x = 15$

$x = 5$; OE = $5(5) + 4 = 29$

26. $x(x - 2) = x + 28$

$x^2 - 2x - x - 28 = 0$

$x^2 - 3x - 28 = 0$

$(x - 7)(x + 4) = 0$

$x = 7$ or $x = -4$. If $x = 7$, then

AC = 7 + 28 = 35; DE = $17\frac{1}{2}$.

If $x = -4$, then AC = 24; DE = 12.

27. Given: $\overline{DC}$ is tangent to circles B and A at points D and C, respectively.

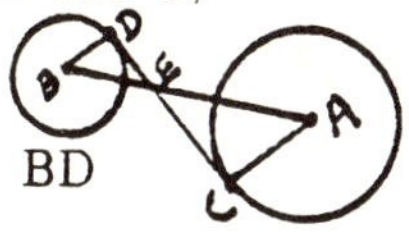

Prove: AC · ED = CE · BD

Proof: If $\overline{DC}$ is tangent to circles B and A at points D and C, then $\overline{BD} \perp \overline{DC}$ and $\overline{AC} \perp \overline{DC}$. ∠s D and C are congruent since they are right angles. ∠DEB ≐ ∠CEA because of vertical angles. △BDE ~ △ACE by AA. It follows that $\frac{AC}{CE} = \frac{BD}{ED}$ since corresponding sides are proportional. Hence, AC · ED = CE · BD.

28. Given: ⊙O with $\overline{EO} \perp \overline{BC}$, $\overline{DO} \perp \overline{BA}$, $\overline{EO} \doteq \overline{OD}$

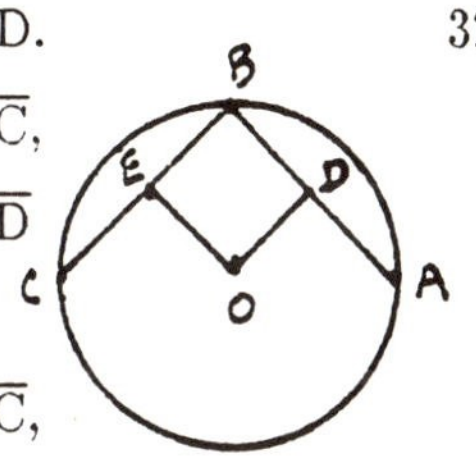

Prove: $\overset{\frown}{BC} \doteq \overset{\frown}{BA}$

Proof: In ⊙O, if $\overline{EO} \perp \overline{BC}$, $\overline{DO} \perp \overline{BA}$, and $\overline{EO} \doteq \overline{OD}$, then $\overline{BC} \doteq \overline{BA}$. (Chords equidistant from the center of the circle are congruent.) It follows then that $\overset{\frown}{BC} \doteq \overset{\frown}{BA}$.

29. Given: $\overline{AP}$ and $\overline{BP}$ are tangent to ⊙Q at A and B; C is the midpoint of $\overset{\frown}{AB}$.

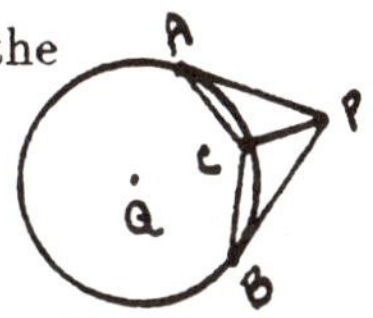

Prove: $\overrightarrow{PC}$ bisects ∠APB

Proof: If $\overline{AP}$ and $\overline{BP}$ are tangent to ⊙Q at A and B, then $\overline{AP} \doteq \overline{BP}$. $\overset{\frown}{AC} \doteq \overset{\frown}{BC}$ since C is the midpoint of $\overset{\frown}{AB}$. It follows that $\overline{AC} \doteq \overline{BC}$ and using $\overline{CP} \doteq \overline{CP}$, we have △ACP ≐ △BCP by SSS. ∠APC ≐ ∠BPC by CPCTC and hence $\overrightarrow{PC}$ bisects ∠APB.

30. If $m\overset{\frown}{AD} = 136$ and $\overline{AC}$ is a diameter, then $m\overset{\frown}{DC} = 44$. If $m\overset{\frown}{BC} = 50$, then $m\overset{\frown}{AB} = 130$.

m∠1 = 93; m∠2 = 25; m∠3 = 43;

m∠4 = 68; m∠5 = 90; m∠6 = 22;

m∠7 = 68; m∠8 = 22; m∠9 = 50;

m∠10 = 112

31. Each side of the square has length $6\sqrt{2}$. Therefore, the perimeter is $24\sqrt{2}$ cm.

32. The perimeter of the triangle is $15 + 5\sqrt{3}$ cm.

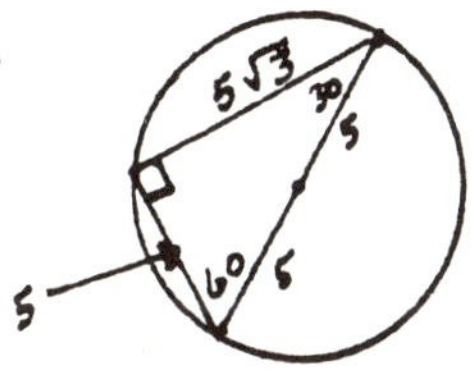

33. $(35 - x)^2 + (6 + x)^2 = 29^2$

$1225 - 70x + x^2 + 36 + 12x + x^2 = 841$

$2x^2 - 58x + 420 = 0$

$x^2 - 29x + 210 = 0$

$(x - 15)(x - 14) = 0$

$x = 15$ or $x = 14$

The lengths of the segments on the hypotenuse are 14 and 15.

34. Let AD = x = AF;

BE = y = DB;

FC = z = CE, then

$$\begin{aligned} x + y \quad\quad &= 9 \\ x \quad\quad + z &= 10 \\ y + z &= 13 \end{aligned}$$

Subtracting the second equation from the first we get $y - z = -1$. Using this one along with the third equation, we have

$$y - z = -1$$
$$y + z = 13.$$

Adding, we get $2y = 12$ or $y = 6$. Solving for x and z, $x = 3$ and $z = 7$. AD = 3; BE = 6; FC = 7.

35. a. AB > CD

b. QP < QR

c. $m\angle A < m\angle C$

36. a.

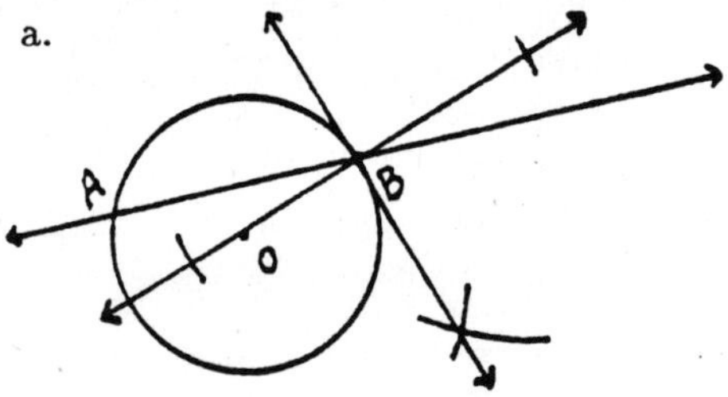

b.

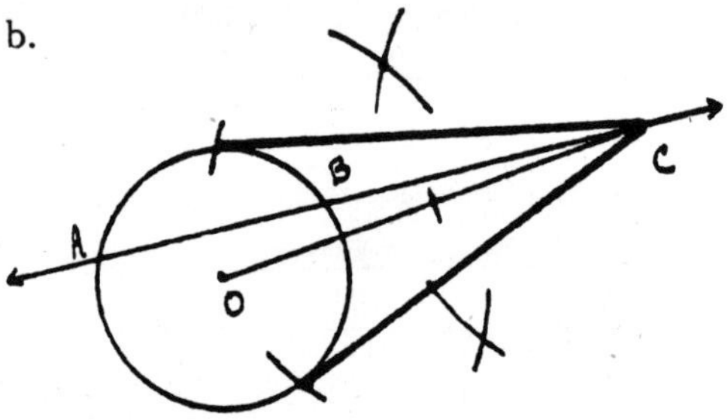

37.

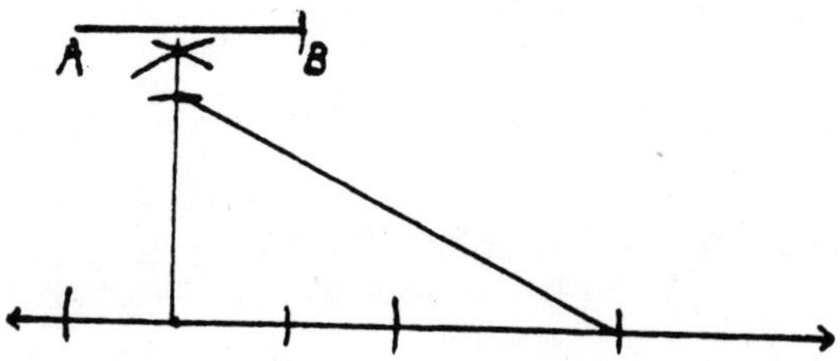

38.

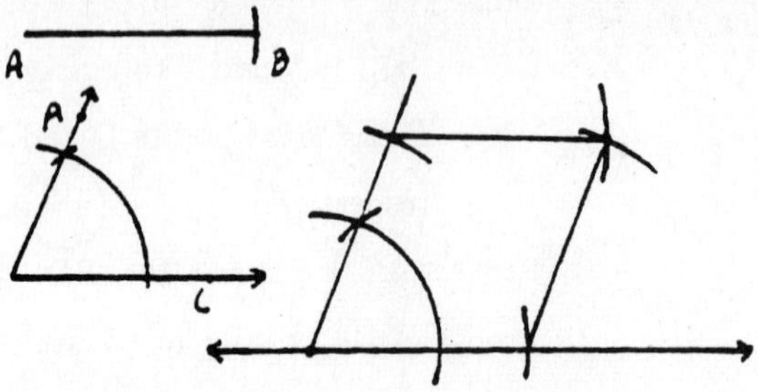

39. The locus of the midpoints of the radii of a circle is another circle.

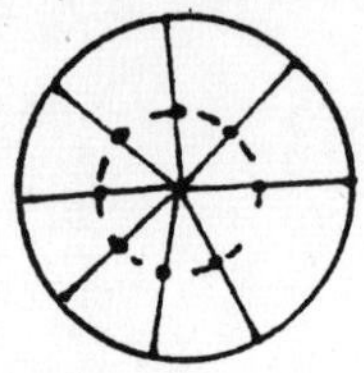

40. The locus of the centers of all circles passing through two given points is the perpendicular bisector of the segment joining the 2 given points.

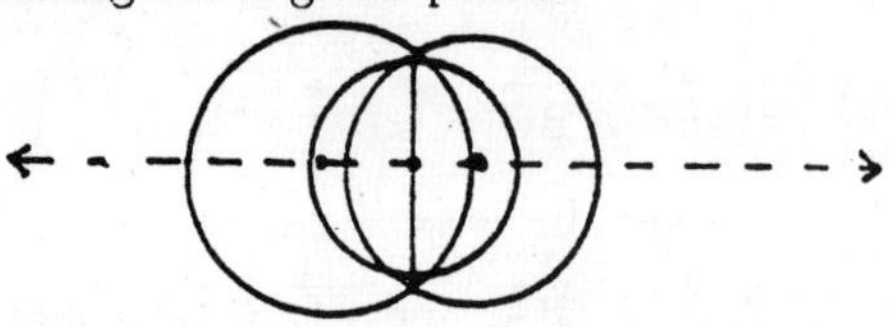

41. The locus of the centers of a penny that rolls around a half-dollar is a circle.

42. The locus of points in space less than three units from a given point is the interior of a sphere.

43. The locus of points equidistant from 2 parallel planes is a parallel plane midway between the 2 planes.

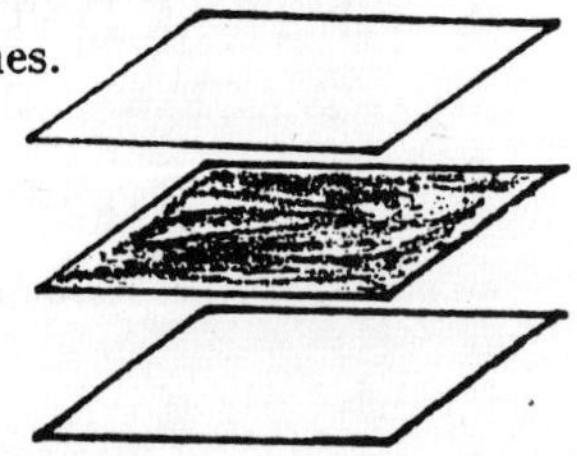

44.

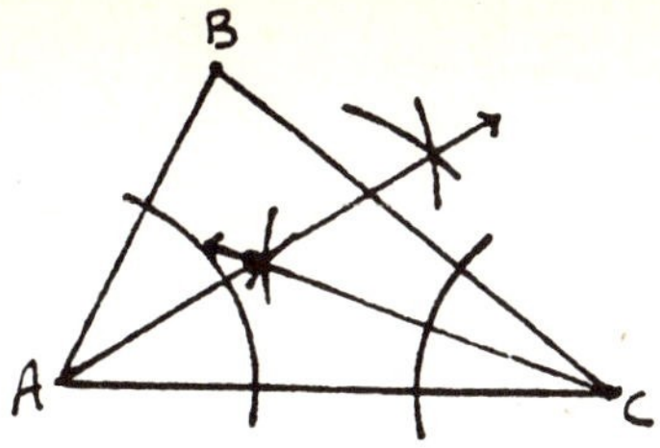

45.

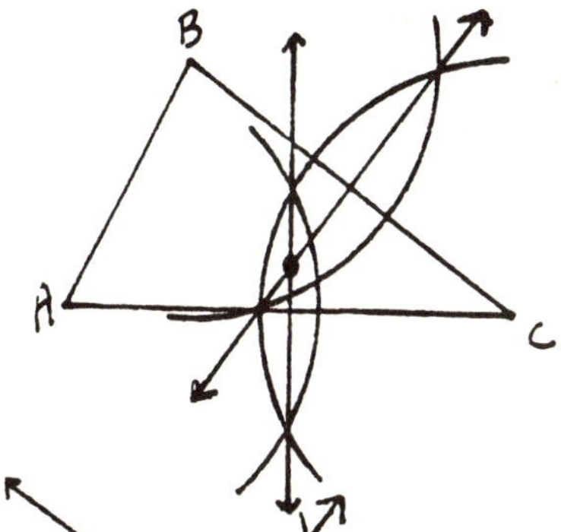

46.

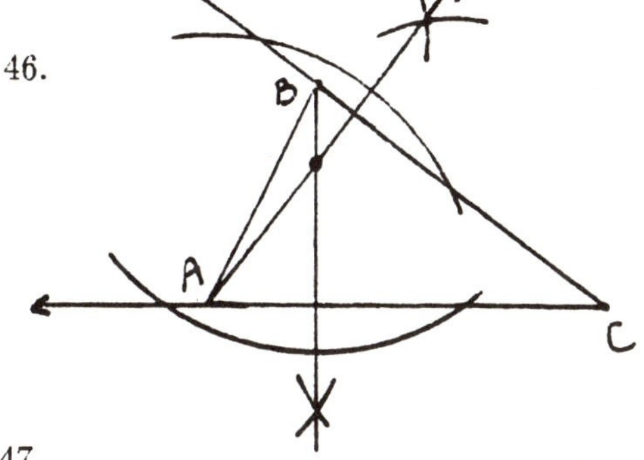

47.

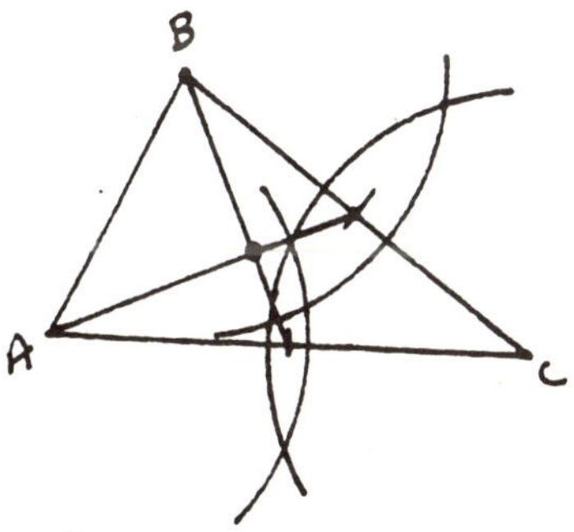

48.

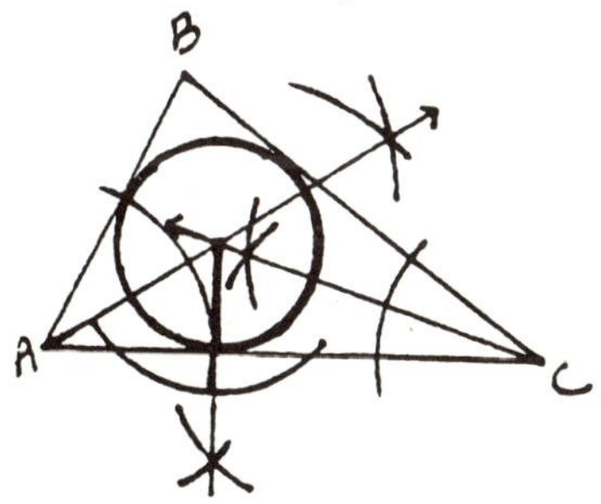

49.

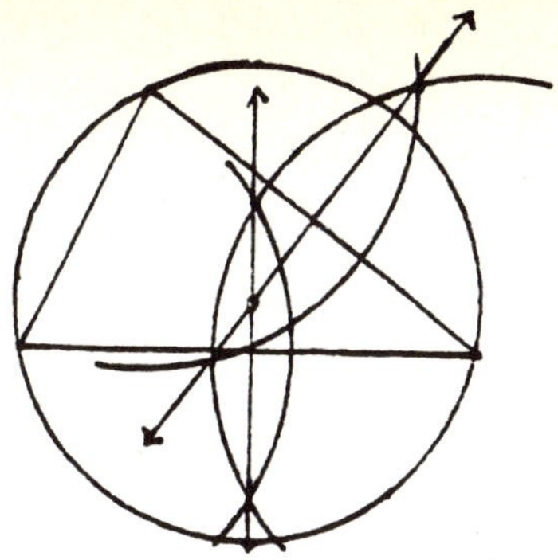

50. a. $BG = \frac{2}{3}(BF) = \frac{2}{3}(18) = 12$

b. $AG = \frac{2}{3}(AE)$

$4 = \frac{2}{3}(AE)$

$AE = \frac{3}{2}(4) = 6$; $GE = 2$.

c. $CG = \frac{2}{3}(DC)$

$4\sqrt{3} = \frac{2}{3}(DC)$

$DC = \frac{3}{2}(4\sqrt{3}) = 6\sqrt{3}$; $DG = 2\sqrt{3}$.

51. $AG = 2(GE)$ and $BG = 2(GF)$

$2x + 2y = 2(2x - y)$ and $3y + 1 = 2(x)$

Simplifying: $2x + 2y = 4x - 2y$ and

$3y + 1 = 2x$

$-2x + 4y = 0$ and $2x - 3y = 1$

Adding the above two equations gives

$y = 1$.

Solving for x gives $-2x + 4 = 0 \rightarrow$

$-2x = -4 \rightarrow x = 2$.

$BF = BG + GF = 4 + 2 = 6$;

$AE = AG + GE = 6 + 3 = 9$.

Chapter Seven: Areas of Polygons and Circles

SECTION 7.1: Area and Initial Postulates

1 . Two triangles with equal areas are not necessarily congruent.

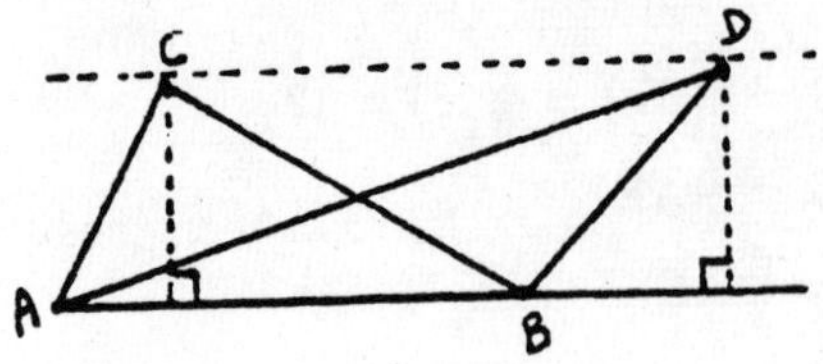

$\triangle ABC \not\cong \triangle ABD$

Two squares with equal areas must be congruent because the sides will be congruent.

5 . The altitudes to $\overline{PN}$ and to $\overline{MN}$ are congruent.

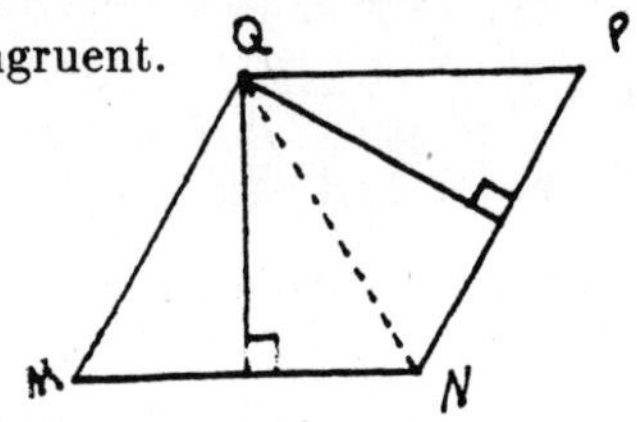

This follows from the fact that $\triangle$s QMN and QPN are congruent. Corresponding altitudes of $\cong$ $\triangle$s are $\cong$.

9 . $A = bh$

$A = 6 \cdot 9$

$A = 54 \text{ cm}^2$

13. $A = bh$

$A = 12 \cdot 6$

$A = 72 \text{ in}^2$

17. Using the Pythagorean Triple, (5,12,13), b = 12.

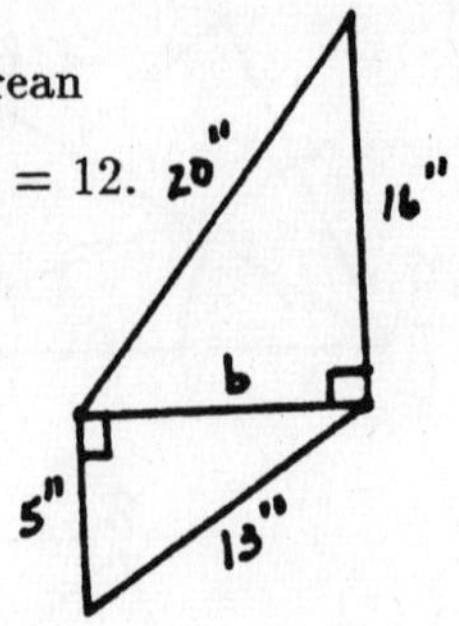

$A = A_{\triangle 1} + A_{\triangle 2}$

$A = \frac{1}{2} \cdot 12 \cdot 16 + \frac{1}{2} \cdot 12 \cdot 5$

$A = 96 + 30$

$A = 126 \text{ in}^2$

21. In the right $\triangle$, h = 8, using the Pythagorean Triple (6,8,10).

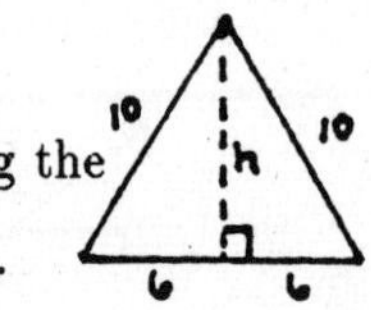

$A = A_{PARALLELOGRAM} + A_{\triangle}$

$A = 12 \cdot 8 + \frac{1}{2} \cdot 12 \cdot 8$

$A = 96 + 48$

$A = 144 \text{ units}^2$

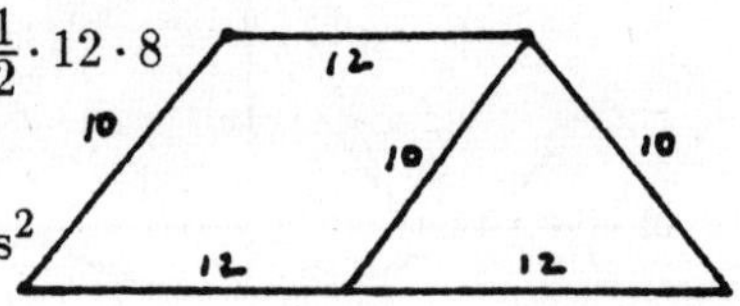

25.

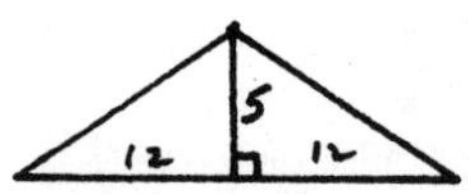

$A = \frac{1}{2}bh + bh$

a. $A = \frac{1}{2} \cdot 24 \cdot 5 + 24 \cdot 10$

$A = 300 \text{ ft}^2$

b. 3 gallons

c. \$46.50

29. a. 3 ft = 1yd ∴ 9 sq ft = 1 sq yd

b. 36 in = 1 yd ∴ 1296 sq in = 1 sq yd

33. Given: Square HJKL with $LJ = d$.

Prove: $A_{HJKL} = \frac{d^2}{2}$

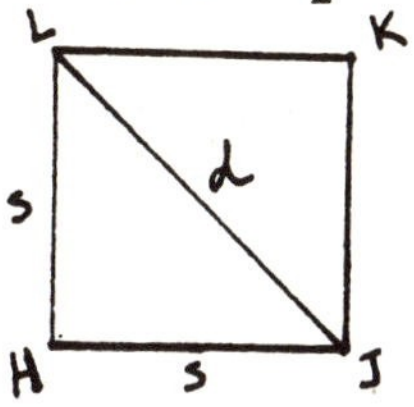

Proof: $A = (LH)(HJ) = s^2$

$s^2 + s^2 = d^2$ by the Pythagorean Theorem.

$2s^2 = d^2$

$s^2 = \frac{d^2}{2}$

$\therefore\ A = \frac{d^2}{2}$

37. a. Let $BD = h$.

$h^2 + x^2 = 13^2$ and

$h^2 + (14 - x)^2 = 15^2$

Then $h^2 = 169 - x^2$, so

$(169 - x^2) + (14 - x)^2 = 225$

$169 - x^2 + 196 - 28x + x^2 = 225$

$-28x + 365 = 225$

$-28x = -140 \rightarrow x = 5.$

Because $h^2 = 169 - x^2$

$h^2 = 169 - 25$

$h^2 = 144 \rightarrow h = 12.$

b. $A_{ABC} = \frac{1}{2}bh \rightarrow A_{ABC} = \frac{1}{2} \cdot 14 \cdot 12$

$\therefore\ A_{ABC} = 84\ in^2.$

41.

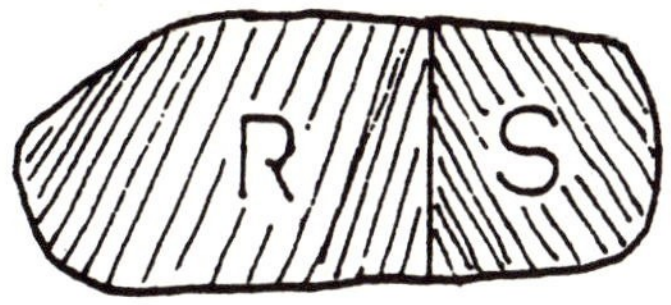

By the Area-Addition Postulate,

$A_{R \cup S} = A_R + A_S$.

Now $A_{R \cup S}$, A_R, and A_S are all positive numbers.

Let p represent the area of region S,

so that $A_{R \cup S} = A_R + p$.

By the definition of inequality, $A_R < A_{R \cup S}$

or $A_{R \cup S} > A_R$.

45. Using $c^2 = a^2 + b^2$,

$c^2 = 5^2 + 12^2 \rightarrow c^2 = 169 \rightarrow c = 13$

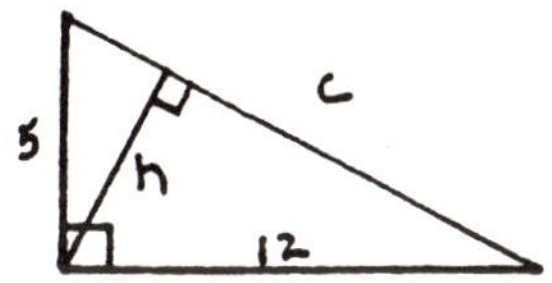

Using the 12″ side as base, the length of its altitude is 5″.

$\therefore\ A = \frac{1}{2} \cdot 12 \cdot 5 \rightarrow A = 30\ in^2$

Also $A = \frac{1}{2}ch$

$A = \frac{1}{2} \cdot 13 \cdot h$

$A = \frac{13}{2}h$

then $\frac{13}{2}h = 30$

Mult. by $\frac{2}{13}$, $\frac{2}{13} \cdot \frac{13}{2}h = \frac{2}{13} \cdot 30$

$h = \frac{60}{13}$ or $4\frac{8}{13}$ in.

49. a. 10

b. 26

c. 18

d. No

SECTION 7.2: Perimeter and Area of Polygons

1. $c = 13$ using the Pythagorean Triple $(5,12,13)$.

$P = a + b + c \rightarrow P = 5 + 12 + 13$

$P = 30$ in.

5. 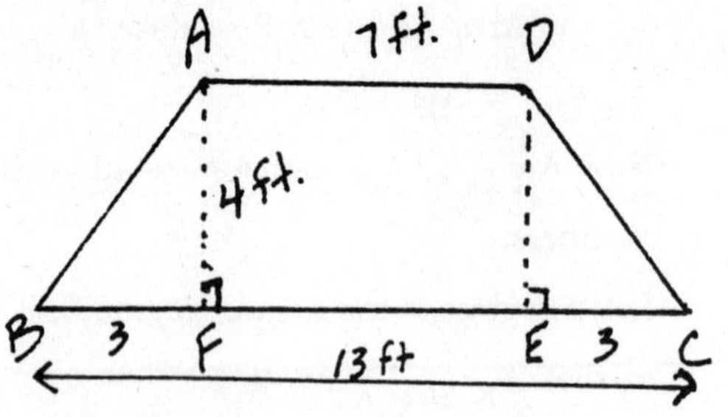

Draw an altitude from D to $\overline{BC}$. $FE = 7$ and $BF = EC = 3$. $AB = DC = 5$. Therefore the perimeter of isosceles trapezoid ADCB with altitudes $\overline{AF}$ and $\overline{DE} = 7 + 5 + 5 + 13 = 30$ feet.

9. $A = \sqrt{s(s - a)(s - b)(s - c)}$ where $a = 13$, $b = 14$, $c = 15$ and

$s = \frac{1}{2}(13 + 14 + 15)$.

$s = \frac{1}{2}(42) = 21 \therefore$

$A = \sqrt{21(21 - 13)(21 - 14)(21 - 15)}$

$A = \sqrt{21(8)(7)(6)}$

$A = \sqrt{(3 \cdot 7) \cdot (2 \cdot 4) \cdot 7 \cdot (2 \cdot 3)}$

$A = \sqrt{2^2 \cdot 3^2 \cdot 2^2 \cdot 7^2}$

$A = 2 \cdot 3 \cdot 2 \cdot 7$

$A = 84 \text{ in}^2$

13. ABCD is a rhombus

$A = \frac{1}{2}d_1 \cdot d_2$

$A = \frac{1}{2}(10)(16)$

$A = 80 \text{ units}^2$

17. Let x = length of second side of triangle.

$2x$ = length of first side of triangle.

$x + 12$ = length of third side of triangle.

$P = a + b + c$

$76 = x + 2x + (x + 12)$

$76 = 4x + 12$

$4x = 64$

$x = 16 \rightarrow 2x = 32 \rightarrow x + 12 = 28$

The lengths of the sides of the triangle are 16 in., 32 in., and 28 in.

21. a. $\frac{A_1}{A_2} = \left(\frac{s_1}{s_2}\right)^2 = \left(\frac{3}{2}\right)^2 = \frac{9}{4}$

b. $\frac{A_1}{A_2} = \left(\frac{s_1}{s_2}\right)^2 = \left(\frac{2}{1}\right)^2 = \frac{4}{1}$ or 4

25. $OA = 5 \rightarrow AC = 10$

$\angle$s B and D are right angles (inscribed in semicircles.)

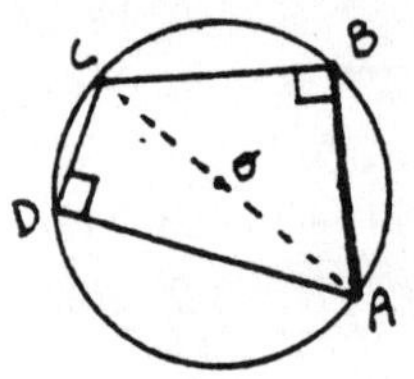

In rt. $\triangle ABC$, $(AC)^2 = (AB)^2 + (BC)^2$.

$10^2 = (AB)^2 + 6^2$

$100 = (AB)^2 + 36 \rightarrow$

$(AB)^2 = 64 \rightarrow AB = 8$.

In rt. $\triangle ADC$, $(AC)^2 = (AD)^2 + (CD)^2$

$10^2 = (AD)^2 + 4^2$

$100 = (AD)^2 + 16$

$84 = (AD)^2 \rightarrow AD = \sqrt{84}$

$= \sqrt{4} \cdot \sqrt{21} = 2\sqrt{21}$.

Now $A_{ABCD} = A_{ABC} + A_{ADC}$

$A_{ABCD} = \frac{1}{2}(6)(8) + \frac{1}{2}(4)(2\sqrt{21})$

$A_{ABCD} = (24 + 4\sqrt{21}) \text{ units}^2$

29. $A = bh$ and $P = 2b + 2h$

$bh = 48$ and $2b + 2h = 28$

$b + h = 14$

$b = 14 - h$

$\therefore$ $bh = 48$ becomes $(14 - h) \cdot h = 48$

$14h - h^2 = 48$

$h^2 - 14h + 48 = 0$

$(h - 6)(h - 8) = 0$

$h - 6 = 0$ or

$h - 8 = 0$

$h = 6$ or $h = 8$

If $h = 6$, then $b = 8$ and vice versa. The dimensions of the garden are 6 yd. by 8 yd.

33. $A = A_{RECT.\#1} + A_{RECT.\#2}$

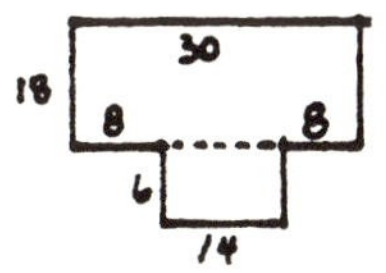

$A = (18)(30) + (6)(14)$

$A = 540 + 84$

$A = 624 \text{ ft}^2$

37. The 4 $\cong$ $\triangle$s have sides of lengths 5, 12, 13.

a. Perimeter = 52 b. Area = 169

41. The area of a trapezoid $= \frac{1}{2}h(b_1 + b_2)$
$= h \cdot \frac{1}{2}(b_1 + b_2) = h \cdot m$ since
median, m, of a trapezoid $= \frac{1}{2}(b_1 + b_2)$

45. $A = \frac{1}{2}d^2 = \frac{1}{2}(\sqrt{10})^2 = \frac{1}{2}(10) = 5 \text{ in}^2$

SECTION 7.3: Regular Polygons and Area

1. First, construct the angle-bisectors of two consecutive angles, say A and B. The point of intersection, O, is the center of the inscribed circle.
Second, construct the line segment $\overline{OM}$ which is perpendicular to $\overline{AB}$. Then, using the radius $r = OM$, construct the inscribed circle with the center O.

5. In $\odot$O, draw diameter $\overline{AB}$.

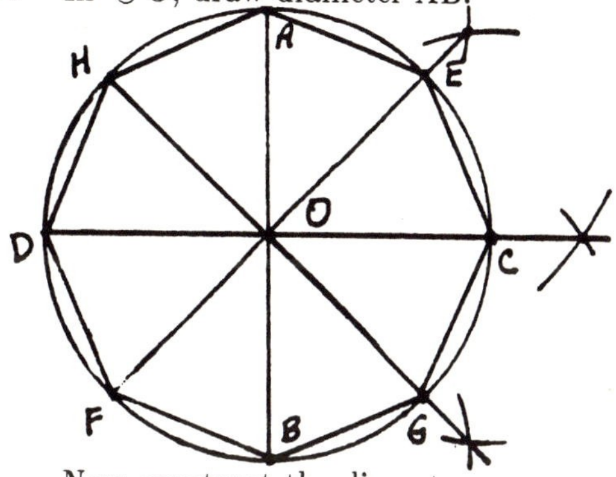

Now construct the diameter $\overline{CD}$ which is $\perp$ $\overline{AB}$ at O. Construct the angle bisectors for $\angle$s AOC and BOC, and extend these to form diameters $\overline{EF}$ and $\overline{GH}$. Joining in the order A, E, C, G, B, F, D, and H determines a regular octagon inscribed in $\odot$O.

9. In square ABCD, apothem $a = 5''$. (Using the 45°-45°-90° relationship). Also radius $r = 5\sqrt{2}''$.

13. $c = \dfrac{360}{n}$

a. 120° b. 90° c. 72° d. 60°

17. $A = \frac{1}{2}aP = \frac{1}{2}(5.2)(37.5) = 97.5 \text{ cm}^2$

21.

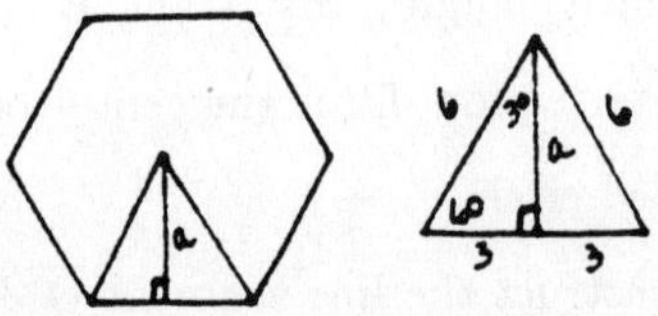

The length of the apothem is $a = 3\sqrt{3}$ (using the 30°-60°-90° relationship).

$A = \frac{1}{2}aP$ becomes

$A = \frac{1}{2} \cdot 3\sqrt{3} \cdot 36$

$A = 54\sqrt{3} \text{ cm}^2$

25. $P = 12(2) = 24$

$A = \frac{1}{2}aP$

$A = \frac{1}{2}(2 + \sqrt{3})24$

$A = 12(2 + \sqrt{3})$

$A = (24 + 12\sqrt{3}) \text{ in}^2$

29.

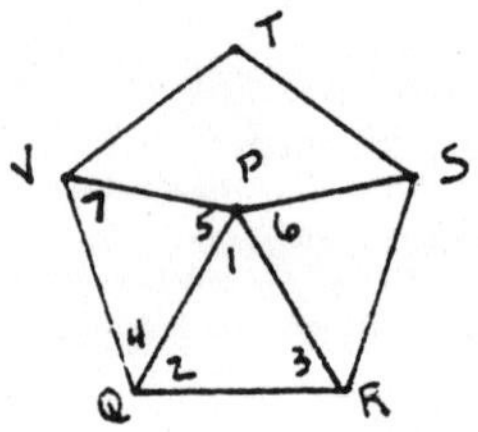

$m\angle 1 = m\angle 2 = m\angle 3 = 60.$

$m\angle VQR = \frac{(5-2)180}{5} = 108;$

So $m\angle 4 = 108 - 60 = 48°$.

With $\overline{VQ} \doteq \overline{QR}$ and $\overline{PQ} \doteq \overline{QR}$,

$\overline{VQ} \doteq \overline{PQ}$, $m\angle 7 = m\angle 5$.

$\therefore\ m\angle 5 = 66°$

Similarly, $m\angle 6 = 66°$. $m\angle VPS =$

$360 - (m\angle 5 + m\angle 1 + m\angle 6)$

$m\angle VPS = 168°$

SECTION 7.4: Circumference and Area of a Circle

1. $C = 2\pi r \qquad A = \pi r^2$

$C = 2 \cdot \pi \cdot 8 \quad A = \pi \cdot 8^2$

$C = 16\pi \text{ cm} \quad A = 64\pi \text{ cm}^2$

5. a. $C = 2\pi r$

$44\pi = 2\pi r$

$\frac{44\pi}{2\pi} = \frac{2\pi r}{2\pi} \rightarrow r = 22 \text{ in.}$

$\therefore\ d = 44 \text{ in.}$

b. $C = 2\pi r$

$60\pi = 2\pi r$

$\frac{60\pi}{2\pi} = \frac{2\pi r}{2\pi} \rightarrow r = 30 \text{ ft.}$

$\therefore\ d = 60 \text{ ft.}$

9. $\ell = \frac{m}{360} \cdot C$

$\ell = \frac{60}{360} \cdot 2 \cdot \pi \cdot 8$

$\ell = \frac{1}{6} \cdot 16\pi$

$\ell = \frac{8}{3}\pi \text{ in}$

13. $A = \pi r^2$

$143 = \pi r^2$

$\frac{143}{\pi} = r^2$

$r \approx 6.7$

17. The maximum area of a rectangle occurs when it is a square.

$\therefore\ A = 16$ sq in.

21. $A = A_{CIRCLE} - A_{SQUARE}$

$A = \pi(4\sqrt{2})^2 - 8^2$

$A = (32\pi - 64) \text{ in}^2$

25. $A = \pi r^2$

$154 = \pi r^2$

$r^2 = \frac{154}{\pi}$

$r \approx 7$ cm

29. Given: Concentric circles with radii of lengths R and r, where R > r.

Explain: $Area_{RING} = \pi(R + r)(R - r)$

$A = A_{LARGER\ CIRCLE} - A_{SMALLER\ CIRCLE}$

$A = \pi R^2 - \pi r^2$

$A = \pi(R^2 - r^2)$.

But $R^2 - r^2$ is a difference of 2 squares, so that $A = \pi(R + r)(R - r)$

33. a. $A = \pi r^2$

$A = \pi(8^2)$

$A \approx 201.06$ ft^2

b. $\frac{201.06}{70} \approx 2.87$ pints

Thus, 3 pints need to be purchased.

c. 3 x \$2.95 = \$8.85

37. $P_{POLYGON} \approx C_{CIRCLE}$

$P \approx 2\pi r$

$P \approx 2 \cdot \pi \cdot 7$

$P \approx 43.98$ cm

41. $C = 2\pi r$

$C = 2\pi(4375)$

$C \approx 27{,}488.94$ miles

SECTION 7.5: More Area Relationships in a Circle

1. $P_{SECTOR} = 10 + 10 + 14 = 34$ in

5. $A_\triangle = \frac{1}{2}rP$

$A_\triangle = \frac{1}{2}r(3s)$

$A_\triangle = \frac{3}{2}rs$

9. $A_\triangle = \frac{1}{2}rP$

$A_\triangle = \frac{1}{2}(2)(6 + 8 + 10)$

$A_\triangle = 24$ sq in

13. $P = 2r + \ell$

$P = 2(8) + \frac{8}{3}\pi$

$P = (16 + \frac{8}{3}\pi)$ in

$A = \frac{m}{360} \cdot \pi r^2$

$A = \frac{60}{360} \cdot \pi(8^2)$

$A = \frac{1}{6} \cdot 64\pi$

$A = \frac{32}{3}\pi$ in^2

17. $P = AB + \ell \overset{\frown}{AB}$

$P = 12 + \frac{60}{360} \cdot 2\pi(12)$

$P = 12 + \frac{1}{6}(24\pi)$

$P = (12 + 4\pi)$ in

$A = A_{SECTOR} - A_\triangle$

$A = \frac{1}{6}(\pi)(12^2) - \frac{12^2}{4}\sqrt{3}$

$A = \frac{1}{6}(144\pi) - \frac{144}{4}\sqrt{3}$

$A = (24\pi - 36\sqrt{3})$ in^2

21. $A = \frac{m}{360} \cdot \pi r^2$

$\frac{9}{4}\pi = \frac{40}{360} \cdot \pi r^2$

$\frac{9}{4}\pi = \frac{1}{9}\pi r^2$

Multiply by 9

$\frac{81}{4}\pi = \pi r^2$

$r^2 = \frac{81}{4}$

$r = \frac{9}{2}$ cm

25. $\ell = \frac{m}{360} \cdot 2\pi r$

$6\pi = \frac{m}{360} \cdot 2\pi(12)$

$6\pi = \frac{m}{360} \cdot 24\pi$

Dividing by 24π, we get

$\frac{1}{4} = \frac{m}{360}$

Multiplying by 360, we get

$m = 90°$.

29. Draw in two radii to consecutive vertices of the square. Then

$r^2 + r^2 = s^2$

$2r^2 = s^2$

$r^2 = \frac{s^2}{2}$

$r = \frac{s}{\sqrt{2}} = \frac{s\sqrt{2}}{2}$

$A = A_{CIRCLE} - A_{SQUARE}$

$A = \pi r^2 - s^2$

$A = \pi\left(\frac{s\sqrt{2}}{2}\right)^2 - s^2$

$A = \pi \cdot \frac{s^2 \cdot 2}{4} - s^2$

$A = \left(\frac{\pi}{2}\right)s^2 - s^2$

33. $A_\triangle = \sqrt{s(s - a)(s - b)(s - c)}$

Also, $A_\triangle = \frac{1}{2}rP$ or

$A_\triangle = \frac{1}{2}r(a + b + c)$; therefore

$\frac{1}{2}r(a + b + c) = \sqrt{s(s - a)(s - b)(s - c)}$

Solve for r.

$r = \frac{2\sqrt{s(s - a)(s - b)(s - c)}}{a + b + c}$

37. $A = \frac{120}{360} \cdot \pi\left(18^2 - 4^2\right)$

$A = \frac{1}{3} \cdot \pi \cdot \left(324 - 16\right)$

$A = \frac{\pi}{3}(308)$

$A = \frac{308\pi}{3} \approx 322.54$ sq in

Chapter 7: Review

1. 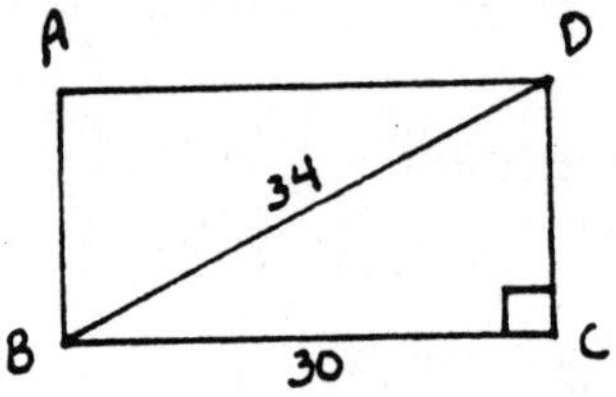

Using the Pythagorean Theeorem,

$(34)^2 = (30)^2 + (DC)^2$

$1156 = 900 + (DC)^2$

$256 = (DC)^2 \rightarrow DC = 16$

$A = 30(16) = 480$ units^2

2. a. $A = 10(4) = 40$ units^2

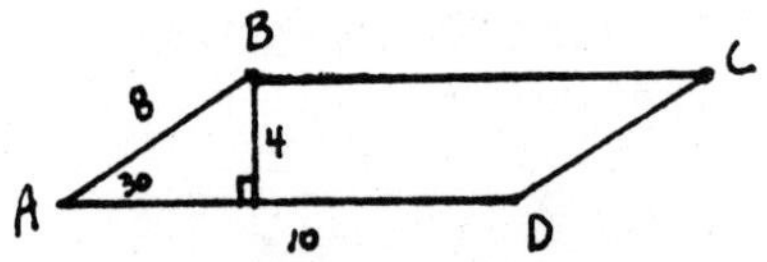

b. $A = 10(4\sqrt{3}) = 40\sqrt{3}$ units^2

c. $A = (10)(4\sqrt{2})$

$A = 40\sqrt{2}$ units^2

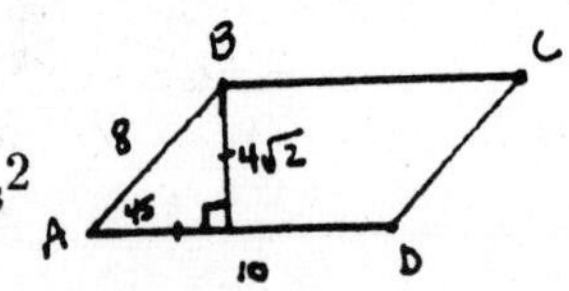

3. Using the 45°-45°-90° relationship,

$AB = DB = 5\sqrt{2}$.

$\therefore\ A_{ABCD} = bh$

$A_{ABCD} = (5\sqrt{2})(5\sqrt{2})$

$A_{ABCD} = 50$ units^2

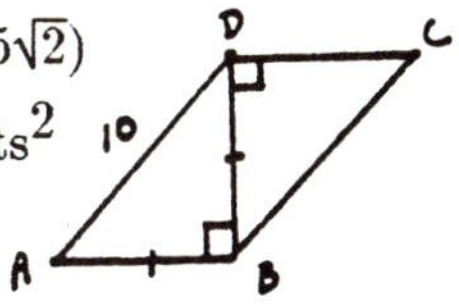

4. $A = \sqrt{s(s-a)(s-b)(s-c)}$

where

$s = \frac{1}{2}(17 + 25 + 26)$

$s = \frac{1}{2}(68)$

$s = 34$

$A = \sqrt{34(34-26)(34-25)(34-17)}$

$A = \sqrt{34(8)(9)(17)}$

$A = \sqrt{2 \cdot 17 \cdot 2 \cdot 4 \cdot 9 \cdot 17}$

$A = \sqrt{2^4 \cdot 3^2 \cdot 17^2}$

$A = 2^2 \cdot 3 \cdot 17$

$A = 204$ units^2

5. $A = \sqrt{s(s-a)(s-b)(s-c)}$

where

$s = \frac{1}{2}(26 + 28 + 30)$

$s = \frac{1}{2}(84) \rightarrow s = 42$

$A = \sqrt{42(42-26)(42-28)(42-30)}$

$A = \sqrt{42(16)(14)(12)}$

$A = \sqrt{2 \cdot 3 \cdot 7 \cdot 2^4 \cdot 2 \cdot 7 \cdot 2^2 \cdot 3}$

$A = \sqrt{2^8 \cdot 3^2 \cdot 7^2}$

$A = 2^4 \cdot 3 \cdot 7$

$A = 16(21) = 336$ units^2

6. Using the (3,4,5) Pythagorean Triple,

$BH = 4$.

$A = \frac{1}{2}h(b_1 + b_2)$

$A = \frac{1}{2} \cdot 4(6 + 12)$

$A = 36$ units^2

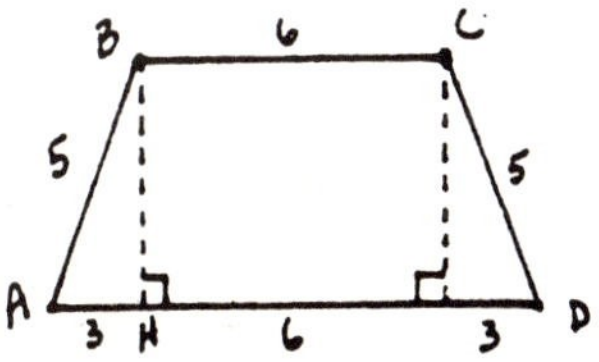

7.

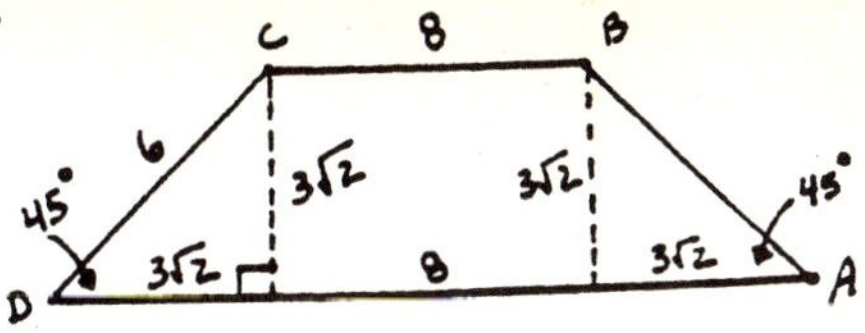

a. $A = \frac{1}{2}h(b_1 + b_2)$

$A = \frac{1}{2} \cdot 3\sqrt{2}[8 + (8 + 6\sqrt{2})]$

$A = \frac{1}{2} \cdot 3\sqrt{2}[16 + 6\sqrt{2}]$

$A = 3\sqrt{2}[8 + 3\sqrt{2}]$

$A = (24\sqrt{2} + 18)$ units^2

b. $A = \frac{1}{2} \cdot 3[8 + (8 + 6\sqrt{3})]$

$A = \frac{3}{2}[16 + 6\sqrt{3}]$

$A = (24 + 9\sqrt{3})$ units^2

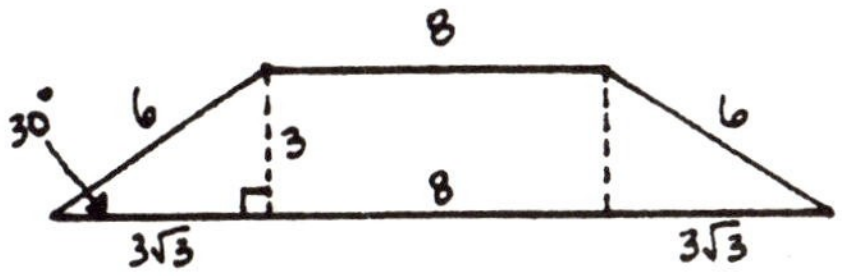

c. $A = \frac{1}{2} \cdot 3\sqrt{3}[8 + 14]$

$A = \frac{1}{2}(3\sqrt{3})(22)$

$A = 33\sqrt{3}$ units^2

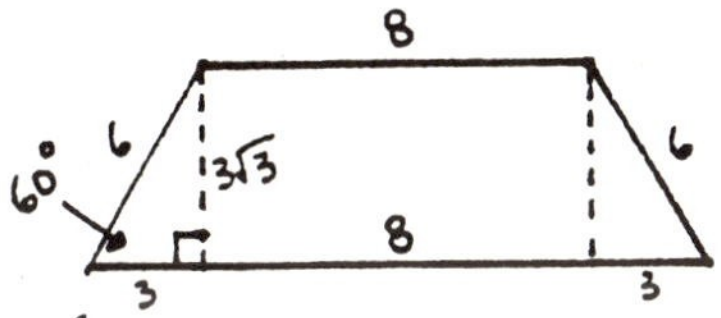

8. $A = \frac{1}{2}d_1 \cdot d_2$

$A = \frac{1}{2}(18)(24)$

$A = 216$ in^2

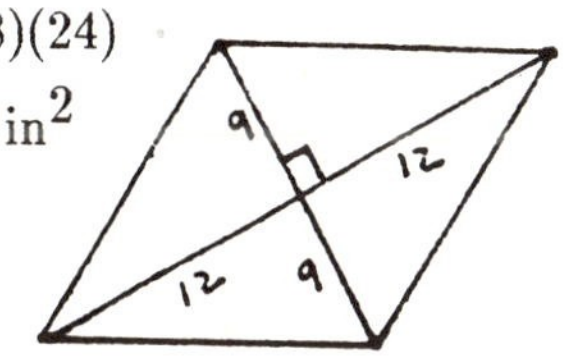

Using the (9,12,15) Pythagorean Triple, $P = 4(15) = 60$ in

9. a. $A = (140)(160) - [(80)(35) + (30)(20)]$

$A = 22{,}400 - [2800 + 600]$

$A = 22{,}400 - 3400$

$A = 19000 \text{ ft}^2$

b. $\frac{19000}{5000} = 3.8$ bags

Tom needs to buy 4 bags

c. Cost $= 4 \cdot \$18 = \$72$

10. a. $A = (9 \cdot 8) + (12 \cdot 8)$

$A = 72 + 96$

$A = 168 \text{ ft}^2$

$\frac{168}{60} = 2.8$ double rolls

3 double rolls are needed

b. $P = 2a + 2b$

$P = 2(9) + 2(12)$

$P = 18 + 24$

$P = 42$ ft. or 14 yd.

$\frac{14}{5} = 2.8$ rolls

3 rolls of border are needed

11. 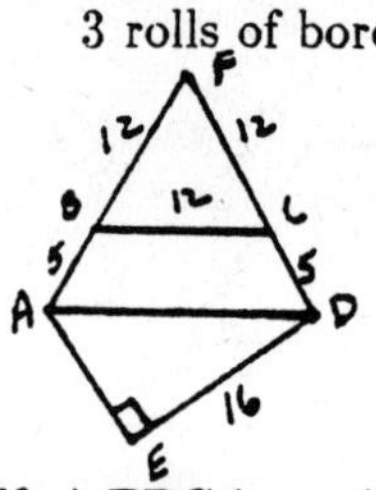

If $\triangle FBC$ is equilateral, then so is $\triangle FAD$.

$\therefore AD = 17$

$(AD)^2 = (AE)^2 + (ED)^2$

$(17)^2 = (AE)^2 + (16)^2$

$289 = (AE)^2 + 256$

$(AE)^2 = 33 \rightarrow AE = \sqrt{33}$

a. $A_{EAFD} = A_{FAD} + A_{AED}$

$A_{EAFD} = \frac{17^2}{4}\sqrt{3} + \frac{1}{2}(16)\sqrt{33}$

$A_{EAFD} = (\frac{289}{4}\sqrt{3} + 8\sqrt{33}) \text{ units}^2$

b. $P_{EAFD} = AF + FD + DE + AE$

$P_{EAFD} = 17 + 17 + 16 + \sqrt{33}$

$P_{EAFD} = (50 + \sqrt{33})$ units

12. 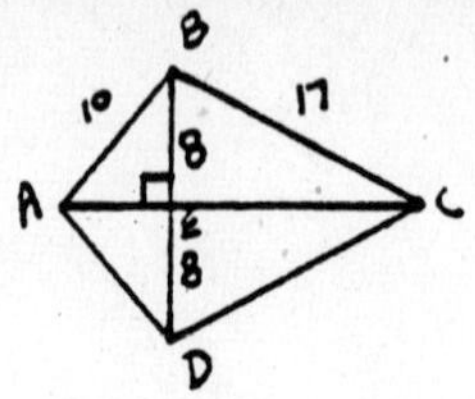

$AE = 6$

(using the 6-8-10 Pythagorean Triple)

$EC = 15$

(using the 8-15-17 Pythagorean Triple)

$A = \frac{1}{2} \cdot d_1 \cdot d_2$

$A = \frac{1}{2}(BD)(AC)$

$A = \frac{1}{2}(16)(6 + 15)$

$A = 8(21)$

$A = 168 \text{ units}^2$

13. 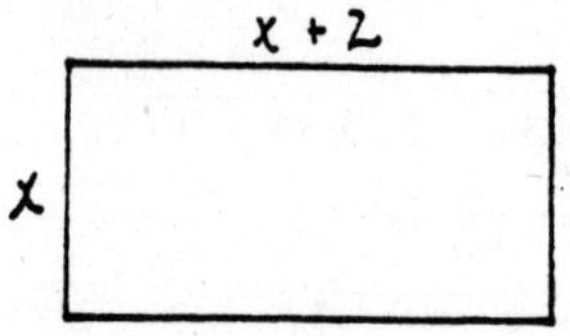

$A = bh$

$35 = x(x + 2)$

$35 = x^2 + 2x$

$x^2 + 2x - 35 = 0$

$(x + 7)(x - 5) = 0$

$x + 7 = 0$ or $x - 5 = 0$

$x = -7$ or $x = 5$; reject -7

The rectangle is 5 cm by 7 cm.

14. $P = a + b + c$

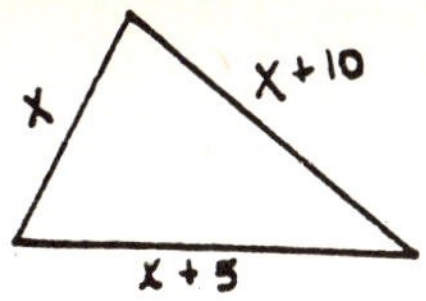

a. $60 = x + (x + 10) + (x + 5)$

$60 = 3x + 15$

$3x = 45$

$x = 15 \rightarrow x + 10 = 25 \rightarrow x + 5 = 20$

The three sides have lengths 15 cm, 25 cm, and 20 cm.

b. (15,20,25) is a Pythagorean Triple in which 25 is the length of the hypotenuse.

$A = \frac{1}{2}bh$

$A = \frac{1}{2}(20)(15)$

$A = 150 \text{ cm}^2$

15.

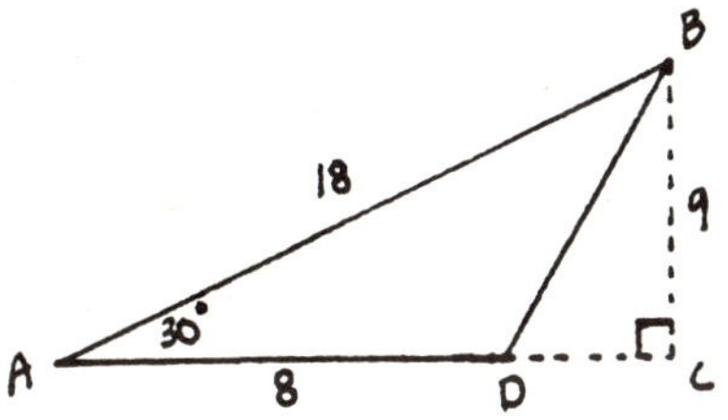

Using the 30°-60°-90° relationship, BC = 9.

$A = \frac{1}{2}bh$

$A = \frac{1}{2}(8)(9)$

$A = 36 \text{ units}^2$

16. $A = \frac{s^2}{4}\sqrt{3}$

$A = \frac{12^2}{4}\sqrt{3}$

$A = \frac{144}{4}\sqrt{3} \rightarrow A = 36\sqrt{3} \text{ cm}^2$

17.

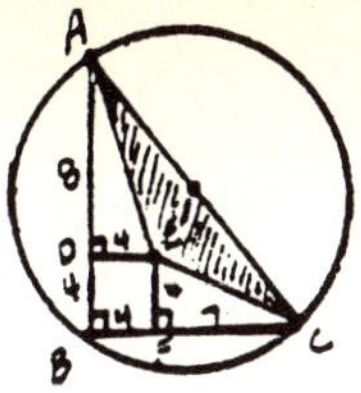

$A = A_{ABC} - (A_{ADE} + A_{BDEF} + A_{EFC})$

$A = \frac{1}{2}(11)(12) - [\frac{1}{2}(8)(4) + 4^2 + \frac{1}{2}(4)(7)]$

$A = 66 - [16 + 16 + 14]$

$A = 66 - 46$

$A = 20 \text{ units}^2$

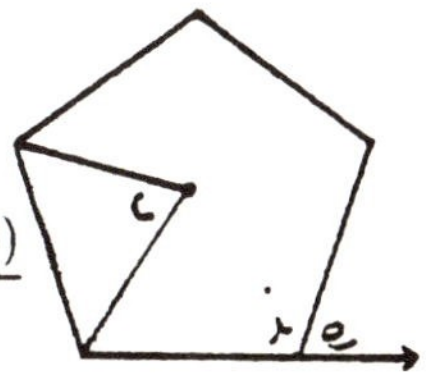

18. a. $c = \frac{360}{5} = 72°$

b. $i = \frac{(5-2)\,180}{5} = \frac{3(180)}{5} = \frac{540}{5} = 108°$

c. $e = 180° - 108° = 72°$

19.

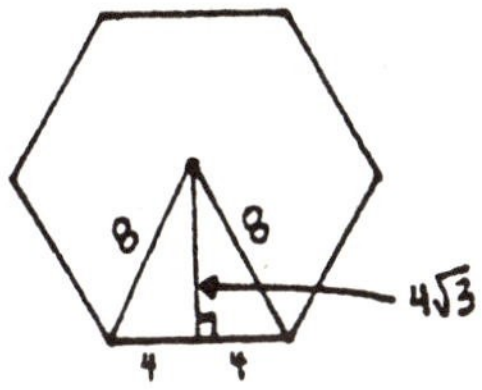

$A = \frac{1}{2}aP$

$A = \frac{1}{2}(4\sqrt{3})(6 \cdot 8)$

$A = 96\sqrt{3} \text{ ft}^2$

20. $P = 3(12\sqrt{3}) = 36\sqrt{3}$

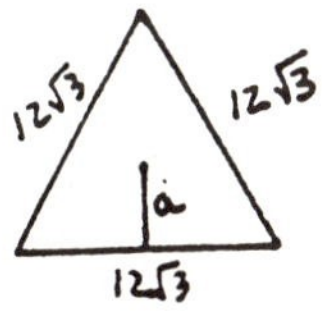

$A = \frac{1}{2}aP$

$108\sqrt{3} = \frac{1}{2} \cdot a \cdot 36\sqrt{3}$

$108\sqrt{3} = (18\sqrt{3})a$

$a = 6$ in.

21. $x \cdot \sqrt{3} = 9$

$x = \frac{9}{\sqrt{3}}$

$x = \frac{9}{\sqrt{3}} \cdot \frac{\sqrt{3}}{\sqrt{3}} = \frac{9\sqrt{3}}{3} = 3\sqrt{3}$

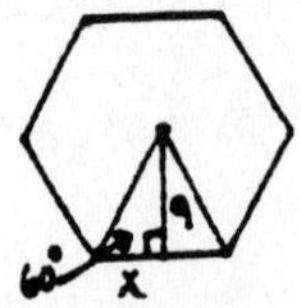

Each side has the length $6\sqrt{3}$ in.

$P = n(s)$

$P = 6(6\sqrt{3})$

$P = 36\sqrt{3}$ inches

$A = \frac{1}{2}aP$

$A = \frac{1}{2}(9)(36\sqrt{3})$

$A = 162\sqrt{3}\ in^2$

22. a. $c = \frac{360}{n}$

$45 = \frac{360}{n}$

$45n = 360$

$n = 8$

b. $P = n(s)$

$P = 8(5)$

$P = 40$

$A = \frac{1}{2}aP$

$A \approx \frac{1}{2}(6)(40)$

$A \approx 120\ cm^2$

23. a. No. ⊥ bisectors of sides of a parallelogram are not necessarily concurrent.

b. No. ⊥ bisectors of sides of a rhombus are not concurrent.

c. Yes. ⊥ bisectors of sides of a rectangle are concurrent.

d. Yes. ⊥ bisectors of sides of a square are concurrent.

24. a. No. ∠bisectors of a parallelogram are not necessarily concurrent.

b. Yes. ∠bisectors of a rhombus are concurrent.

c. No. ∠bisectors of a rectangle are not necessarily concurrent.

d. Yes. ∠bisectors of a square are concurrent.

25. If the radius of the inscribed circle is 7 in. then the length of each side of the triangle is $14\sqrt{3}$ in.

$A_{\triangle} = \frac{1}{2}rP$

$A_{\triangle} = \frac{1}{2}(7)(3 \cdot 14\sqrt{3})$

$A_{\triangle} = 147\sqrt{3}$

$A_{\triangle} \approx 254.61$ sq in.

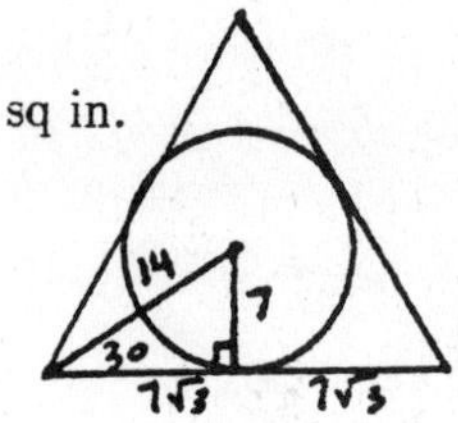

26. a. $A = (20)(30) - (12)(24)$

$A = 600 - 288$

$A = 312\ ft^2$

b. $\frac{312}{9} = 34\frac{2}{3}\ yd^2$

$35\ yd^2$ should be purchased

c. 35 x \$9.97 = \$348.95

27. $A = A_{SQUARE} - 4 \cdot A_{SECTOR}$

$A = 8^2 - 4 \cdot \left[\frac{90}{360}(\pi \cdot 4^2)\right]$

$A = 64 - 4\left[\frac{1}{4}(16\pi r)\right]$

$A = (64 - 16\pi)$ units^2

28. $A = A_{SEMICIRCLE} - A_{TRIANGLE}$

$A = \frac{1}{2}(\pi \cdot 7^2) - \frac{1}{2}(7)(7\sqrt{3})$

$A = (\frac{49}{2}\pi - \frac{49}{2}\sqrt{3})$ units^2

29. $A_{SEGMENT} = A_{SECTOR} - A_{TRIANGLE}$

$$A = \frac{60}{360} \cdot \pi \cdot 4^2 - \frac{4^2}{4}\sqrt{3}$$

$$A = \left(\frac{8}{3}\pi - 4\sqrt{3}\right) \text{units}^2$$

30.

$A = A_{RECT} - 2 \cdot A_{CIRCLE}$

$A = (12)(24) - 2 \cdot (\pi \cdot 6^2)$

$A = (288 - 72\pi)\ \text{units}^2$

31.

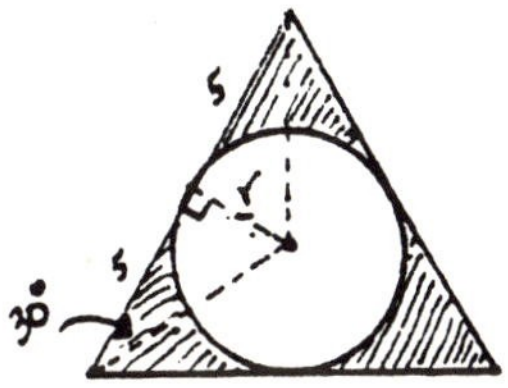

The radius of the circle is r.

$r\sqrt{3} = 5$

$r = \frac{5}{\sqrt{3}} = \frac{5\sqrt{3}}{3}$

$A = A_{EQ.\,\Delta} - A_{CIRCLE}$

$$A = \frac{10^2}{4}\sqrt{3} - \pi\left(\frac{5\sqrt{3}}{3}\right)^2$$

$$A = 25\sqrt{3} - \pi \cdot \frac{75}{9}$$

$$A = \left(25\sqrt{3} - \frac{25}{3}\pi\right) \text{units}^2$$

32.

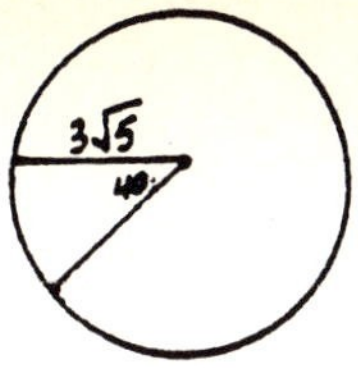

$$\ell = \frac{m}{360}(2\pi r)$$

$$\ell = \frac{40}{360}(2 \cdot \pi \cdot 3\sqrt{5})$$

$$\ell = \frac{1}{9}(6\pi\sqrt{5})$$

$$\ell = \frac{2\pi\sqrt{5}}{3}\ \text{cm}$$

$$A = \frac{m}{360}(\pi r^2)$$

$$A = \frac{40}{360}\left(\pi \cdot (3\sqrt{5})^2\right)$$

$$A = \frac{1}{9}(\pi \cdot 45)$$

$A = 5\pi\ \text{cm}^2$

33. a. $C = \pi d$

$66 = \frac{22}{7} \cdot d$

$\frac{7}{22} \cdot 66 = \frac{7}{22} \cdot \frac{22}{7}d$

$d = 21$ ft.

b. $d = 21 \rightarrow r = \frac{21}{2}$

$A = \pi r^2$

$A = \frac{22}{7} \cdot \left(\frac{21}{2}\right)^2$

$A = \frac{693}{2} \approx 346\frac{1}{2}\ \text{ft}^2$

34. a. $A_{SECTOR} = \frac{m}{360}(\pi r^2)$

$A = \frac{80}{360}(27\pi)$

$A = \frac{2}{9}(27\pi)$

$A = 6\pi \text{ ft}^2$

b. Since $\pi r^2 = 27\pi$

$r^2 = 27$

$r = \sqrt{27} = \sqrt{9} \cdot \sqrt{3} = 3\sqrt{3}$

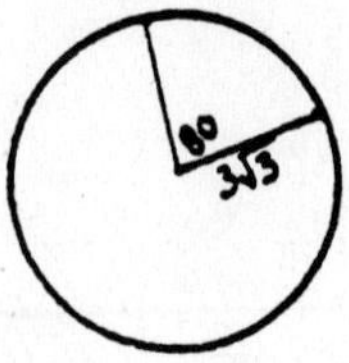

$\ell = \frac{m}{360}(2\pi r)$

$\ell = \frac{80}{360}(2\pi \cdot 3\sqrt{3})$

$\ell = \frac{2}{9}(6\pi\sqrt{3})$

$\ell = \frac{4\pi}{3}\sqrt{3}$

$P = 2r + \ell$

$P = 2(3\sqrt{3}) + \frac{4\pi}{3}\sqrt{3}$

$P = \left(6\sqrt{3} + \frac{4\pi}{3}\sqrt{3}\right)$ ft.

35. 

$A_{SHADED} = A_{SEMICIRCLE} - A_{\triangle}$

$A_{SHADED} = \frac{1}{2} \cdot (\pi \cdot 6^2) - \frac{1}{2} \cdot 6 \cdot 12$

$A_{SHADED} = 18\pi - 36$

The area sought is one-half the shaded area.

$A = \frac{1}{2}(18\pi - 36)$

$A = (9\pi - 18) \text{ in}^2$

36. Given: Concentric circles with radii of lengths R and r with R > r; O is the center of the circles.

Prove: $A_{RING} = \pi(BC)^2$

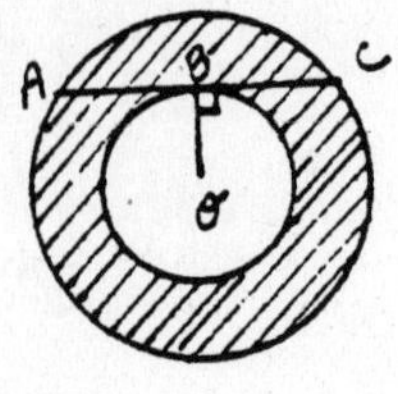

Proof: By an earlier theorem,

$A_{RING} = \pi R^2 - \pi r^2$

$= \pi(OC)^2 - \pi(OB)^2$

$= \pi[(OC)^2 - (OB)^2]$

In rt. $\triangle OBC$,

$(OB)^2 + (BC)^2 = (OC)^2$

$\therefore (OC)^2 - (OB)^2 = (BC)^2$

In turn, $A_{RING} = \pi(BC)^2$.

37. The area of a circle circumscribed about a square is twice the area of the circle inscribed within the square.

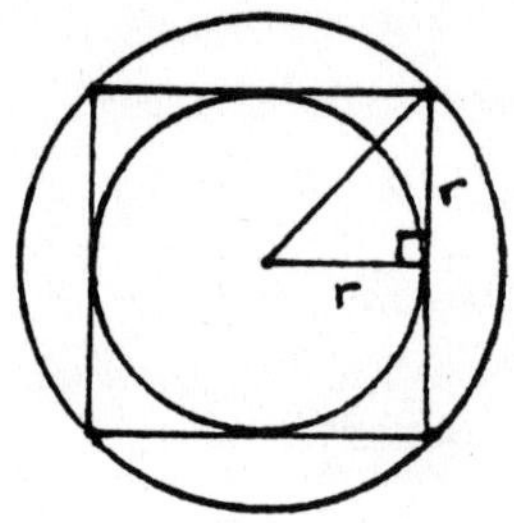

Proof: Let r represent the length of radius of the inscribed circle.

Using the 45°-45°-90° relationship, the radius of the larger circle is $r\sqrt{2}$.

Now $A_{INSCRIBED\ CIRCLE} = \pi r^2$

$A_{CIRCUMSCRIBED\ CIRCLE} = \pi(r\sqrt{2})^2$

$= \pi \cdot (r^2 \cdot 2)$

$= 2(\pi r^2)$

$= 2\left(A_{INSCRIBED\ CIRCLE}\right)$

38. If semicircles are constructed on each of the sides of a right triangle, then the area of the semicircle on the hypotenuse is equal to the sum of the areas of the semicircles on the two legs.

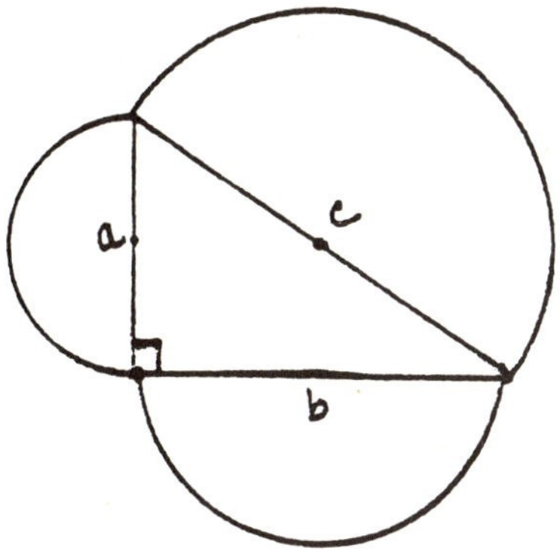

Proof: The radii of the semicircles are $\frac{1}{2}a$, $\frac{1}{2}b$, and $\frac{1}{2}c$. The area of the semicircle on the hypotenuse is

$A = \frac{1}{2}[\pi(\frac{1}{2}c)^2]$

$= \frac{1}{2} \cdot \pi(\frac{1}{4}c^2)$

$= \frac{1}{2}\pi[\frac{1}{4}(a^2 + b^2)]$ ($c^2 = a^2 + b^2$ by the Pythagorean Theorem.)

$= \frac{1}{2}\pi[(\frac{1}{4}a^2 + \frac{1}{4}b^2)$

$= \frac{1}{2}\pi\left(\frac{1}{4}a^2\right) + \frac{1}{2}\pi\left(\frac{1}{4}b^2\right)$

$= \frac{1}{2}\pi\left(\frac{1}{2}a\right)^2 + \frac{1}{2}\pi\left(\frac{1}{2}b\right)^2$

$= \frac{1}{2}[\pi\left(\frac{1}{2}a\right)^2] + \frac{1}{2}[\pi\left(\frac{1}{2}b\right)^2]$,

which is the sum of areas of the semicircles on the two legs.

39. a. $A = (18)(15) - [\frac{1}{2}(3.14)3^2 + \frac{1}{4}(3.14) \cdot 3^2]$

$A = 270 - 21.2$

$A = 248.8\ ft^2$

The approx. number of yd^2 of carpeting needed is $\frac{248.8}{9} \approx 28$.

b. From (a), the number of ft^2 to be tiled is $21.2\ ft^2$.

40. a. $A = \frac{1}{2}[\pi R^2 - \pi r^2]$

$A = \frac{1}{2}[(3.14)(30)^2 - (3.14)(18)^2]$

$A = \frac{1}{2}[1808.64]$

$A = 904.32\ ft^2 \approx 905\ ft^2$

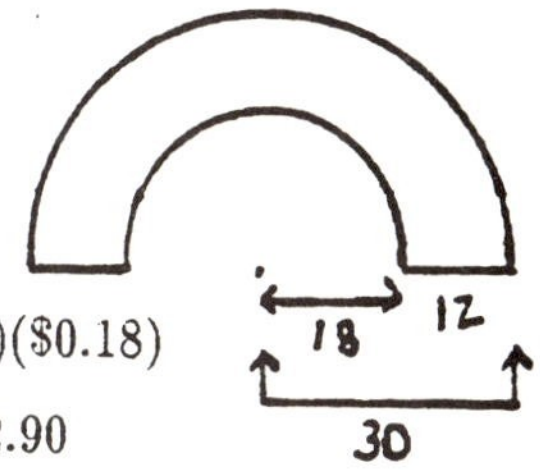

b. Cost = (905)($0.18)

Cost = $162.90

c. Length $= \frac{1}{2}(2\pi R) + \frac{1}{2}(2\pi r)$

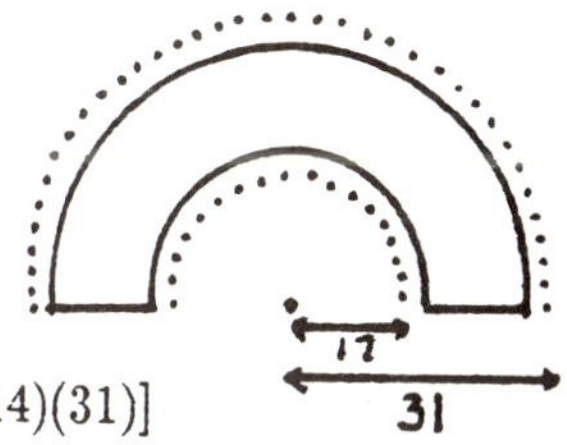

$= \frac{1}{2}[2(3.14)(31)]$

$+ \frac{1}{2}[2(3.14)(17)]$

$= 97.34 + 52.38$

$= 150.72$

Approximately 151 flowers are needed.

Chapter Eight: Surfaces and Solids

SECTION 8.1: Prisms, Area and Volume

1. a. Yes
   b. Oblique
   c. Hexagon
   d. Oblique hexagonal prism
   e. Parallelogram

5. a. cm^2
   b. cm^3

9. $V = Bh$
   $V = 12(10)$
   $V = 120\ cm^3$

13. a. 2n
    b. n
    c. 2n
    d. 3n
    e. n
    f. 2
    g. n + 2

17. a. $L = hP$
    $L = (6)(3 + 4 + 5)$
    $L = 72\ ft^2$
    b. $T = L + 2B$
    $T = 72 + 2(\frac{1}{2}bh)$
    $T = 72 + 4 \cdot 3$
    $T = 84\ ft^2$
    c. $V = Bh$
    $V = (\frac{1}{2}bh) \cdot 6$
    $V = (\frac{1}{2} \cdot 4 \cdot 3)(6)$
    $V = 36\ ft^3$

21. $T = 4x + 4(x + 2) + x(x + 2) + 4x + 4(x + 2) + x(x + 2)$
    $94 = 4x + 4x + 8 + x^2 + 2x + 4x + 4x + 8 + x^2 + 2x$
    $94 = 2x^2 + 20x + 16$
    $0 = 2x^2 + 20x - 78$
    $0 = 2(x^2 + 10x - 39)$
    $0 = 2(x + 13)(x - 3)$
    $x + 13 = 0$ or $x - 3 = 0$
    $x = -13$ or $x = 3$; reject $x = -13$

25. V = Volume of rectangular prism + Volume of triangular prism
    $V = 8 \cdot 10 \cdot 7 + \frac{1}{2} \cdot 8 \cdot 2 \cdot 10$
    $V = 560 + 80$
    $V = 640\ ft^3$

29. The diagonal of a base is $e\sqrt{2}$.
    Let the diagonal of the cube be D.
    $D = \sqrt{(e\sqrt{2})^2 + e^2}$
    $D = \sqrt{e^2 \cdot 2 + e^2}$
    $D = \sqrt{3e^2}$
    $D = e\sqrt{3}$

33. $V = 12 \cdot 12 \cdot 6 = 864\ in^3$

37. $L = 5 \cdot A_{PARALLELOGRAM}$
    $L = 5 \cdot bh$
    $L = 5 \cdot 12 \cdot 12$
    $L = 720\ cm^2$

SECTION 8.2: Pyramids, Area and Volume

1. a. Right pentagonal prism
   b. Oblique pentagonal prism

5 . a. Pyramid

b. E

c. $\overline{EA}$, $\overline{EB}$, $\overline{EC}$, $\overline{ED}$

d. $\triangle EAB$, $\triangle EBC$, $\triangle ECD$, $\triangle EAD$

e. No

9 . $T = 12 + 16 + 12 + 10 + 16$

$T = 66 \text{ in}^2$

13. a. $n + 1$

b. n

c. n

d. $2n$

e. n

f. $n + 1$

17. a. $B = \frac{1}{2}aP$

$B = \frac{1}{2}(6.3)(5 \cdot 9.2)$

$B = 144.9 \text{ cm}^2$

b. $V = \frac{1}{3}Bh$

$V = \frac{1}{3}(144.9)(14.6)$

$V = 705.18 \text{ cm}^3$

21. $V = \frac{1}{3}Bh$

$72 = \frac{1}{3} \cdot x^2 \cdot x$

$72 = \frac{1}{3}x^3$

$x^3 = 216$

$x = 6$

$\ell^2 = a^2 + h^2$

$\ell^2 = 3^2 + 6^2$

$\ell^2 = 45$

$\ell = \sqrt{45} = 3\sqrt{5}$

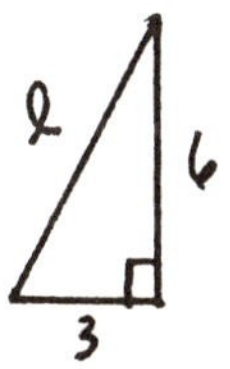

$L = \frac{1}{2}\ell P$

$L = \frac{1}{2}(3\sqrt{5})(4 \cdot 6)$

$L = 36\sqrt{5}$

$T = L + B$

$T = 36\sqrt{5} + (6 \cdot 6)$

$T = 36\sqrt{5} + 36 \approx 116.5 \text{ in}^2$

25. $V = \frac{1}{3}Bh$

$V = \frac{1}{3}(\frac{1}{2}aP)(15)$

$V = \frac{1}{3} \cdot \frac{1}{2} \cdot (7.5)(12 \cdot 4)(15)$

$V = 900 \text{ ft}^3$

29. $V = V_{\text{TALLER PYRAMID}} - V_{\text{SHORTER PYRAMID}}$

$V = \frac{1}{3}B_1h_1 - \frac{1}{3}B_2h_2$

$V = \frac{1}{3}(6 \cdot 6)(32) - \frac{1}{3}(3 \cdot 3)(16)$

$V = 384 - 48$

$V = 336 \text{ in}^3$

33. $T = B + L$

$T = 36 + \frac{1}{2} \cdot 24 \cdot 5$

$T = 36 + 60$

$T = 96 \text{ in}^2$

$\ell^2 + 9 = 34$

$\ell^2 = 25$

$\ell = 5$

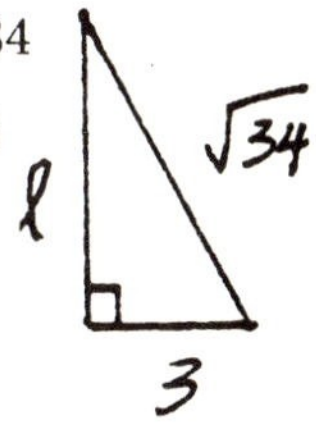

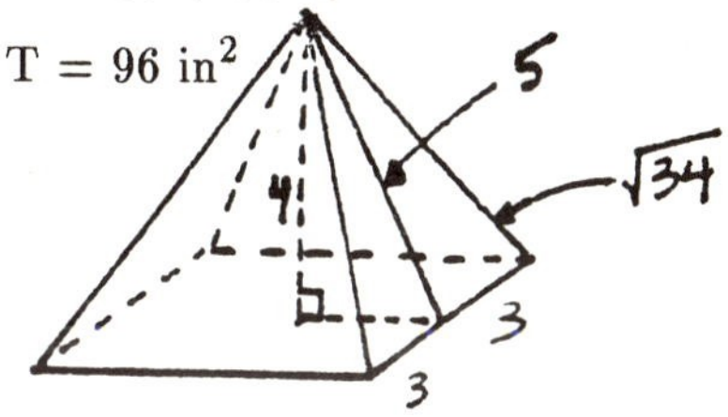

SECTION 8.3: Cylinders and Cones

1 . a. $L = 2\pi rh$

$L = 2\pi \cdot 5 \cdot 6$

$L = 60\pi \approx 188.5 \text{ in}^2$

b. $T = L + 2B$

$T = 60\pi + 2\pi \cdot 5^2$

$T = 60\pi + 50\pi$

$T = 110\pi \approx 345.58 \text{ in}^2$

c. $V = \pi r^2h$

$V = \pi \cdot 5^2 \cdot 6$

$V = 150\pi \approx 471.24 \text{ in}^3$

5. $V = \pi r^2 h$

$200\pi = \pi \cdot r^2 \cdot 8$

$r^2 = \frac{200\pi}{8\pi} = 25$

$r = 5$ cm

9. Using the 45°-45°-90° relationship,

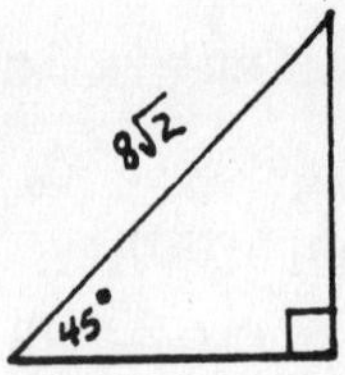

$h\sqrt{2} = 8\sqrt{2}$

$h = \frac{8\sqrt{2}}{\sqrt{2}} \rightarrow h = 8$ in.

$V = \pi r^2 h$

$V \doteq \pi \cdot 2^2 \cdot 8$

$V = 32\pi \approx 100.53$ in^3

13. $r^2 + h^2 = \ell^2$

$(4.8)^2 + h^2 = (5.2)^2$

$h^2 = (5.2)^2 - (4.8)^2$

$h^2 = 4$

$h = 2$ m

17. Let x = the length of the axis of the cone.

$x^2 = 3^2 + 6^2$

$x^2 = 45$

$x = \sqrt{45} = 3\sqrt{5} \approx 6.71$ cm

21. The solid formed is a cylinder with

$r = 3$ and $h = 6$.

$V = Bh$

$V = \pi r^2 h$

$V = \pi \cdot 3^2 \cdot 6$

$V = 54\pi$ in^3

25. The solid formed consists of two cones sharing the same base with $r = 12$. The height of one cone is 16 and the height of the other cone is 9.

$V = \frac{1}{3}Bh_1 + \frac{1}{3}Bh_2$

$V = \frac{1}{3}\pi r^2 h_1 + \frac{1}{3}\pi r^2 h_2$

$V = \frac{1}{3}\pi r^2 (h_1 + h_2)$

$V = \frac{1}{3} \cdot \pi \cdot 12^2 \cdot (16 + 9)$

$V = \frac{1}{3} \cdot \pi \cdot 12^2 \cdot 25$

$V = 1200\pi$ cm^3

29. $V = V_{CYLINDER} - V_{CONE}$

$V = Bh - \frac{1}{3}Bh$

$V = \frac{2}{3}Bh$

$V = \frac{2}{3} \cdot \pi r^2 \cdot h$

$V = \frac{2}{3} \cdot \pi \cdot 6^2 \cdot 8$

$V = 192\pi \approx 603.19$ in^3

33. Let the length of the radius of the base = 3x and let the length of the slant height = 5x.

$\ell^2 = r^2 + h^2$

$(5x)^2 = (3x)^2 + h^2$

$25x^2 = 9x^2 + h^2$

$h^2 = 16x^2$

$h = 4x$

$V = \frac{1}{3}Bh$

$96\pi = \frac{1}{3} \cdot \pi r^2 \cdot 4x$

$96\pi = \frac{1}{3} \cdot \pi \cdot (3x)^2 \cdot 4x$

$96\pi = 12\pi x^3$

$x^3 = 8$

$x = 2$

$r = 3x = 3(2) = 6$ in

$\ell = 5x = 5(2) = 10$ in

$L = \frac{1}{2}\ell C$

$L = \frac{1}{2} \cdot 10 \cdot 2\pi r$

$L = \frac{1}{2} \cdot 10 \cdot 2\pi \cdot 6$

$L = 60\pi \approx 188.5 \text{ in}^2$

37. $V = Bh$

$V = \pi r^2 h$

$V = \pi \cdot 2^2 \cdot 5$

$V = 20\pi$

Capacity $= 20\pi \cdot 7.5 \approx 471.24$ gallons.

41. $V = V_{\text{LARGER CONE}} - V_{\text{SMALLER CONE}}$

$V = \frac{1}{3}\pi R^2 H - \frac{1}{3}\pi r^2 h$

$V = \frac{1}{3}\pi(4^2)(30) - \frac{1}{3}\pi(3^2)(22.5)$

$V \approx 290.60 \text{ cm}^3$

SECTION 8.4: Polyhedrons and Spheres

1. Polyhedron EFGHIJK is concave.

5. A regular hexahedron has 6 faces (F), 8 vertices (V), and 12 edges (E).

$V + F = E + 2$; $8 + 6 = 12 + 2$

9.

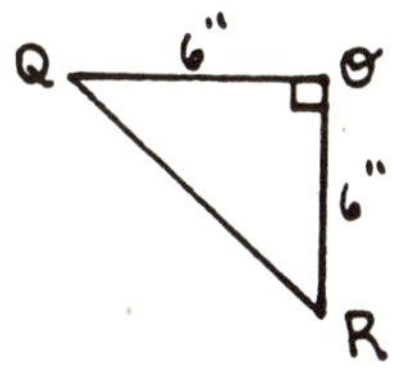

a. Using the 45°-45°-90° relationship, $QR = 6\sqrt{2} \approx 8.49$ in.

b.

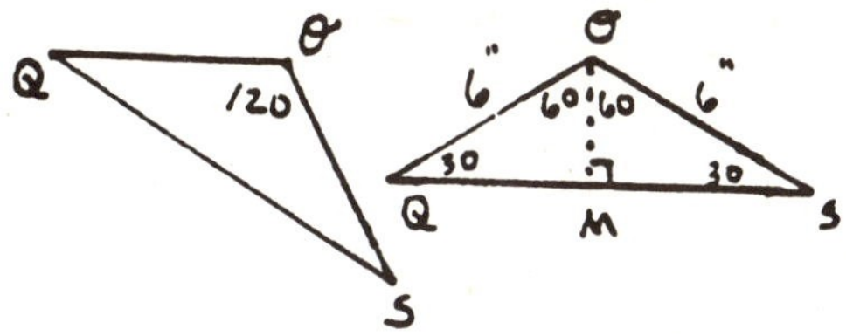

Where M is the midpoint of $\overline{QS}$, $OM = 3$ and $QM = 3\sqrt{3}$ by the 30°-60°-90° relationship.

$QS = 2(QM) = 2(3\sqrt{3}) = 6\sqrt{3}$.

That is, $QS = 6\sqrt{3} \approx 10.39$ in.

13. $S = 6(4.2)^2 = 105.84 \text{ cm}^2$

17. a. $A = 34.9(12) + 52.5(20) = 1468.8 \text{ cm}^2$

b. Cost = $1468.8(\$0.006) = \$8.81$

21.

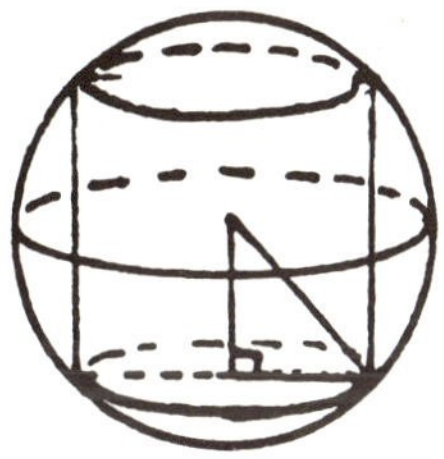

$h = d$ or

$h = 2r$

In the triangle shown, $\frac{h}{2} = r$.

Then $r^2 + r^2 = 6^2$

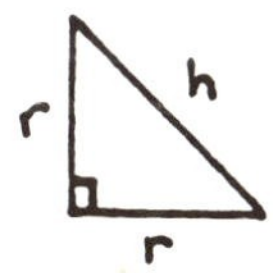

$2r^2 = 36$

$r^2 = 18$

$r = \sqrt{18} = 3\sqrt{2} \approx 4.24$ in

$h = 6\sqrt{2} \approx 8.49$ in

25. a. $S = 4\pi r^2$

$S = 4\pi \cdot 3^2$

$S = 36\pi \approx 113.1 \text{ m}^2$

b. $V = \frac{4}{3}\pi r^3$

$V = \frac{4}{3}\pi \cdot 3^3$

$V = 36\pi \approx 113.1 \text{ m}^3$

29. $S = 4\pi r^2$

$S = 4\pi \cdot 3^2$

$S = 36\pi \approx 113.1 \text{ ft}^2$

The number of pints of paint needed to paint the tank is $\frac{113.1}{40} = 2.836$ pints.

3 pints of paint would have to be purchased.

33. The solid formed is a sphere.

$S = 4\pi r^2$

$S = 4\pi \cdot 3^2$

$S = 36\pi \text{ units}^2$

$V = \frac{4}{3}\pi r^3$

$V = \frac{4}{3}\pi \cdot 3^3$

$V = 36\pi \text{ units}^3$

37. $V = V_{\text{LARGER SPHERE}} - V_{\text{SMALLER SPHERE}}$

$\therefore V = \frac{4}{3}\pi R^3 - \frac{4}{3}\pi r^3$

CHAPTER 9: Review

1. $L = hP$

$L = 12(8 \cdot 7)$

$L = 672 \text{ in}^2$

2. $L = hP$

$L = 11(7 + 8 + 12)$

$L = 297 \text{ cm}^2$

3.

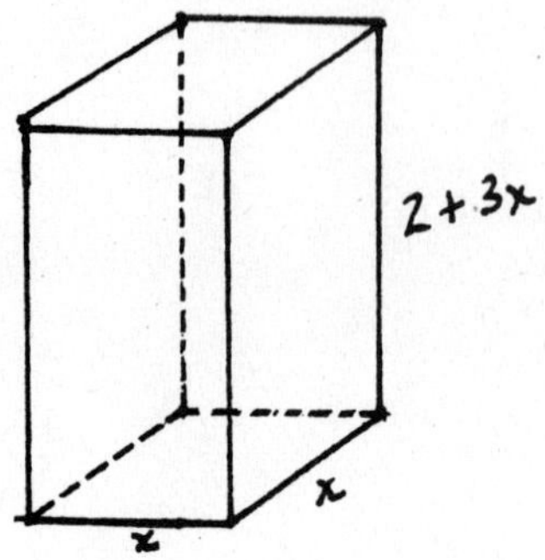

$L = hP$

$480 = (2 + 3x) \cdot 4x$

$480 = 8x + 12x^2$

$0 = 12x^2 + 8x - 480$

$0 = 3x^2 + 2x - 120$

$0 = (3x + 20)(x - 6)$

$3x + 20 = 0$ or $x - 6 = 0$

$x = -\frac{20}{3}$ or $x = 6$; reject $x = -\frac{20}{3}$

Dimensions are 6 in by 6 in by 20 in.

$V = Bh$

$V = (6 \cdot 6)(20)$

$V = 720 \text{ in}^3$

4. 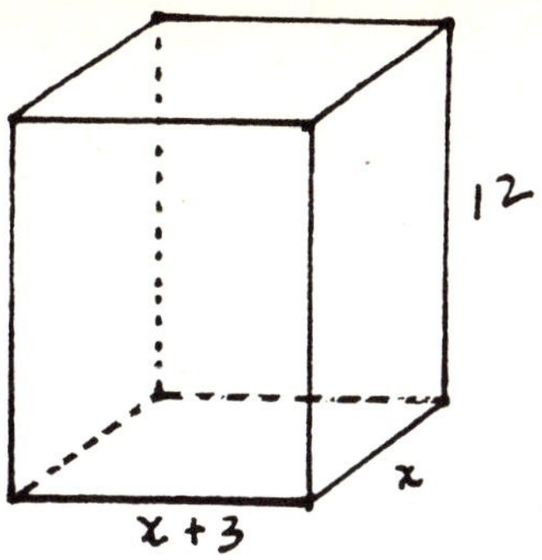

$L = 12(4x + 6)$

$360 = 48x + 72$

$48x = 288$

$x = 6$

The dimensions of the box are

$\ell = 9$ cm, $w = 6$ cm, $h = 12$ cm

$T = L + 2B$

$T = 360 + 2(9 \cdot 6)$

$T = 360 + 108$

$T = 468 \text{ cm}^2$

$V = \ell \cdot w \cdot h$

$V = 9 \cdot 6 \cdot 12$

$V = 648 \text{ cm}^3$

5. The base of the prism is a right triangle since $15^2 = 9^2 + 12^2$.

a. $L = hP$

$L = 10(9 + 12 + 15)$

$L = 360 \text{ in}^2$

b. The area of the base is

$B = \frac{1}{2}bh$

$B = \frac{1}{2} \cdot 12 \cdot 9$

$B = 54$

$T = L + 2B$

$T = 360 + 2 \cdot 54$

$T = 468 \text{ in}^2$

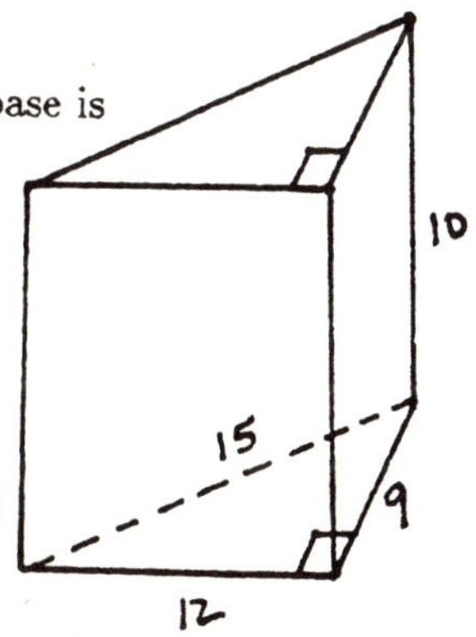

c. $V = Bh$

$V = 54 \cdot 10$

$V = 540 \text{ in}^3$

6. a. $L = hP$

$L = 13(6 \cdot 8)$

$L = 624 \text{ cm}^2$

b. Using the 30°-60°-90° relationship, the apothem of the base is $4\sqrt{3}$.

The area of the base is then equal to

$B = \frac{1}{2}aP$

$B = \frac{1}{2} \cdot 4\sqrt{3} \cdot (6 \cdot 8)$

$B = 96\sqrt{3}$

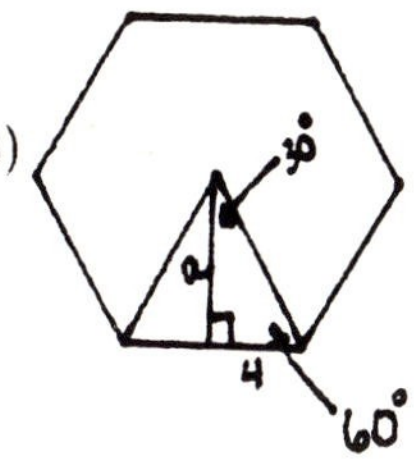

$T = L + 2B$

$T = 624 + 2 \cdot 96\sqrt{3}$

$T = 624 + 192\sqrt{3} \approx 956.55 \text{ cm}^2$

c. $V = Bh$

$V = (96\sqrt{3}) \cdot 13$

$V = 1248\sqrt{3} \approx 2161.6 \text{ cm}^3$

7. $\ell^2 = a^2 + h^2$

$\ell^2 = 5^2 + 8^2$

$\ell^2 = 89$

$\ell = \sqrt{89} \approx 9.43$ cm

8. $\ell^2 = a^2 + h^2$

$12^2 = 9^2 + h^2$

$h^2 = 144 - 81$

$h^2 = 63$

$h = \sqrt{63} = 3\sqrt{7} \approx 7.94$ in

9. $\ell^2 = r^2 + h^2$

$\ell^2 = 5^2 + 7^2$

$\ell^2 = 74$

$\ell = \sqrt{74} \approx 8.60$ in

10. $\ell^2 = r^2 + h^2$

$(2r)^2 = r^2 + 6^2$

$4r^2 = r^2 + 6^2$

$3r^2 = 36$

$r^2 = 12$

$r = \sqrt{12} = 2\sqrt{3} \approx 3.46$ cm

11. a. $L = \frac{1}{2}\ell P$

$L = \frac{1}{2} \cdot 15 \cdot (4 \cdot 18)$

$L = 540$ in^2

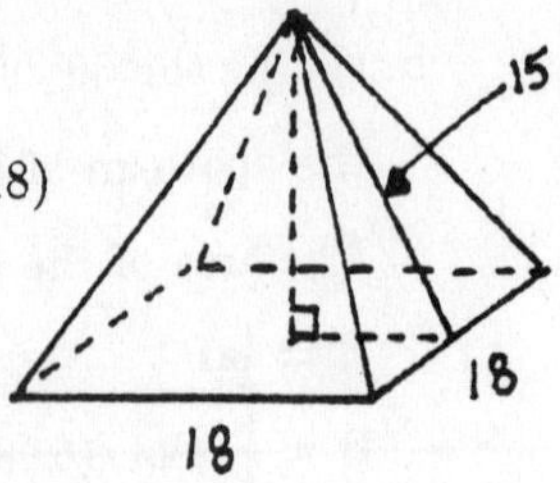

b. $T = L + B$

$T = 540 + 18^2$

$T = 540 + 324$

$T = 864$ in^2

c. $\ell^2 = a^2 + h^2$

$15^2 = 9^2 + h^2$

$h^2 = 225 - 81$

$h^2 = 144$

$h = 12$

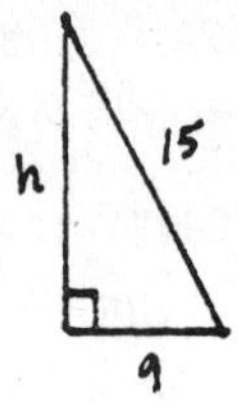

$V = \frac{1}{3}Bh$

$V = \frac{1}{3} \cdot 18^2 \cdot 12$

$V = 1296$ in^3

12. a. Using the 30°-60°-90° relationship, the apothem is $a = 2\sqrt{3}$ cm.

$\ell^2 = a^2 + h^2$

$\ell^2 = (2\sqrt{3})^2 + 8^2$

$\ell^2 = 12 + 64$

$\ell^2 = 76$

$\ell = \sqrt{76} = 2\sqrt{19}$

$L = \frac{1}{2}\ell P$

$L = \frac{1}{2} \cdot 2\sqrt{19} \cdot (12 \cdot 3)$

$L = 36\sqrt{19} \approx 156.92$ cm^2

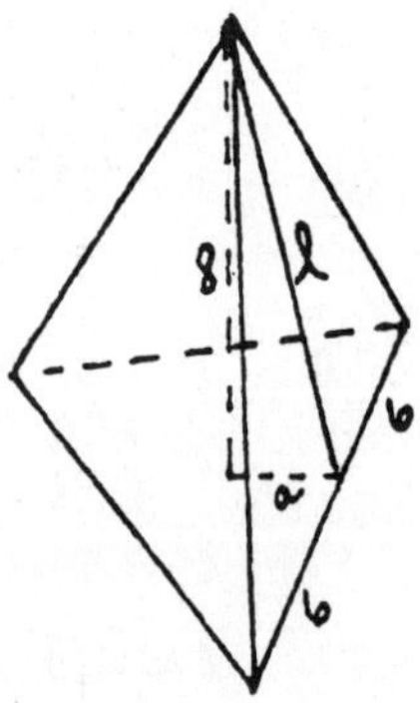

b. $T = L + B$

$T = 36\sqrt{19} + \frac{s^2\sqrt{3}}{4}$

$T = 36\sqrt{19} + \frac{12^2\sqrt{3}}{4}$

$T = 36\sqrt{19} + 36\sqrt{3} \approx 219.27$ cm^2

c. $V = \frac{1}{3}Bh$

$V = \frac{1}{3}(36\sqrt{3})(8)$

$V = 96\sqrt{3} \approx 166.28$ cm^3

13. a. $L = hC$

$L = h \cdot 2\pi r$

$L = 10 \cdot 2\pi \cdot 6$

$L = 120\pi$ in^2

b. $T = L + 2B$

$T = 120\pi + 2\pi r^2$

$T = 120\pi + 2\pi \cdot 6^2$

$T = 120\pi + 72\pi$

$T = 192\pi$ in^2

c. $V = Bh$

$V = \pi r^2 h$

$V = \pi \cdot 6^2 \cdot 10$

$V = 360\pi$ in^3

14. a. $V = \frac{1}{2}Bh$

$V = \frac{1}{2}\pi r^2 h$

$V \approx \frac{1}{2} \cdot 3.14 \cdot 4^2 \cdot 14$

$V \approx 351.68$ ft^3

b. The inside area and outside area represent the total area of the cylinder.

$T = L + 2B$

$T \approx 2(3.14)(4)(14) + 2(3.14) \cdot 4^2$

$T \approx 351.68 + 100.48$

$T \approx 452.16$ ft^2

15. a. $L = \frac{1}{2}\ell C$

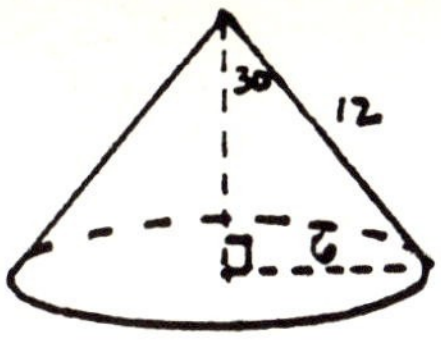

$L = \frac{1}{2}\ell \cdot 2\pi r$

$L = \ell \cdot \pi r$

$L = 12 \cdot \pi \cdot 6$

$L = 72\pi \approx 226.19 \text{ cm}^2$

b. $T = L + B$

$T = 72\pi + \pi r^2$

$T = 72\pi + \pi \cdot 6^2$

$T = 72\pi + 36\pi$

$T = 108\pi \approx 339.29 \text{ cm}^2$

c. $V = \frac{1}{3}Bh$

$V = \frac{1}{3} \cdot \pi r^2 \cdot h$

$V = \frac{1}{3} \cdot \pi \cdot 6^2 \cdot 6\sqrt{3}$

$V = 72\pi\sqrt{3} \approx 391.78 \text{ cm}^3$

16. $V = \frac{1}{3}Bh$

$V = \frac{1}{3}\pi r^2 \cdot h$

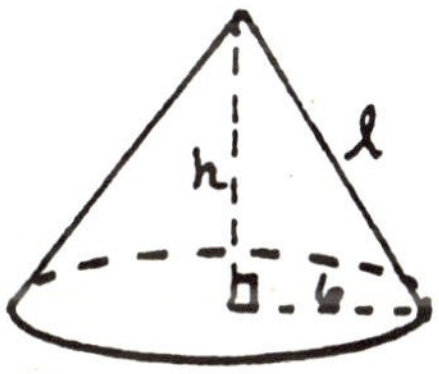

$96\pi = \frac{1}{3}\pi \cdot 6^2 \cdot h$

$96\pi = 12\pi h$

$h = \frac{96\pi}{12\pi} = 8$

$\ell = 10$ using the Pythagorean Triple (6,8,10).

17. $S = 4\pi r^2$

$S \approx 4 \cdot \frac{22}{7} \cdot 7^2$

$S \approx 616 \text{ in}^2$

18. $V = \frac{4}{3}\pi r^3$

$V \approx \frac{4}{3}(3.14) \cdot 6^3$

$V \approx 904.32 \text{ cm}^3$

19.

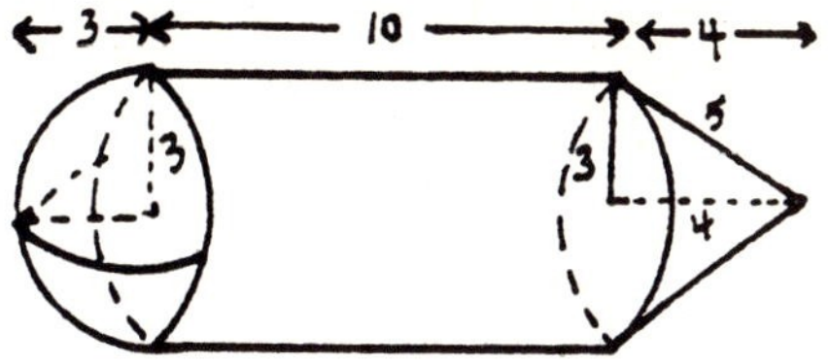

$V = V_{\text{HEMISPHERE}} + V_{\text{CYLINDER}} + V_{\text{CONE}}$

$V = \frac{1}{2}\left(\frac{4}{3}\pi r^3\right) + Bh + \frac{1}{3}Bh$

$V = \frac{1}{2}\left(\frac{4}{3}\pi r^3\right) + \pi r^2 h + \frac{1}{3}\pi r^2 h$

$V = \frac{1}{2}\left(\frac{4}{3}\pi \cdot 3^3\right) + \pi \cdot 3^2 \cdot 10 + \frac{1}{3}\pi \cdot 3^2 \cdot 4$

$V = 18\pi + 90\pi + 12\pi$

$V = 120\pi \text{ units}^3$

20. Let r equal the radius of the smaller sphere while 3r is the radius of the larger sphere.

$$\frac{\text{Surface Area of Smaller}}{\text{Surface Area of Larger}} = \frac{4\pi r^2}{4\pi(3r)^2} = \frac{r^2}{9r^2} = \frac{1}{9}$$

$$\frac{\text{Volume of Smaller}}{\text{Volume of Larger}} = \frac{\frac{4}{3}\pi r^3}{\frac{4}{3}\pi(3r)^3} = \frac{r^3}{27r^3} = \frac{1}{27}$$

21. The solid formed is a cone.

$V = \frac{1}{3}Bh$

$V = \frac{1}{3}\pi r^2 \cdot h$

$V \approx \frac{1}{3} \cdot \frac{22}{7} \cdot 5^2 \cdot 7$

$V \approx 183\frac{1}{3} \text{ in}^3$

22. The solid formed is a cylinder.

$V = Bh$

$V = \pi r^2 h$

$V = \pi \cdot 6^2 \cdot 8$

$V = 288\pi \text{ cm}^3$

23. The solid formed is a sphere.

$V = \frac{4}{3}\pi r^3$

$V = \frac{4}{3}\pi \cdot 2^3$

$V = \frac{32\pi}{3} \text{ in}^3$

24.

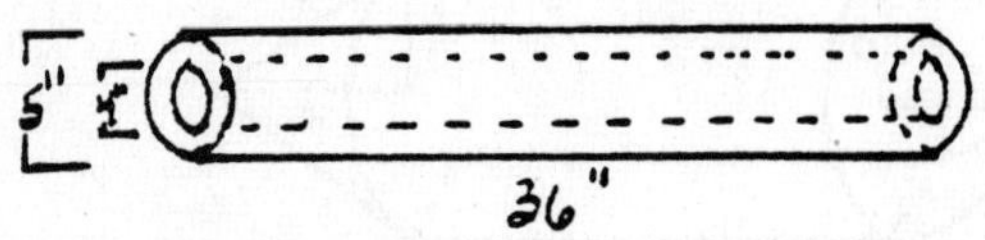

$V = V_{\text{LARGER CYLINDER}} - V_{\text{SMALLER CYLINDER}}$

$V = \pi R^2h - \pi r^2h$

$V \approx (3.14)(5^2)(36) - (3.14)(4^2)(36)$

$V \approx 2826 - 1808.64$

$V \approx 1017.36 \text{ in}^3$

25. $V = V_{\text{CUBE}} - V_{\text{SPHERE}}$

$V = e^3 - \frac{4}{3}\pi r^3$

$V = 14^3 - \frac{4}{3}\pi \cdot 7^3$

$V = 2744 - \frac{1372\pi}{3} \text{ in}^3$

26. a. An octahedron has eight faces which are equilateral triangles.

b. A tetrahedron has four faces which are equilateral triangles.

c. A dodecahedron has twelve faces which are regular pentagons.

27. $V = V_{\text{CYLINDER}} + V_{\text{2 HEMISPHERES}} - V_{\text{SPHERE}}$

Because the volume of the 2 hemispheres equals the volume of a sphere, we have

$V_{\text{PILL}} = V_{\text{CYLINDER}}$

$V = Bh$

$V = \pi r^2h$

$V = \pi \cdot 2^2 \cdot 10$

$V = 40\pi \text{ mm}^3$

28. a. $V = 16 \quad E = 24 \quad F = 10$

$V + F = E + 2$

$16 + 10 = 24 + 2$

b. $V = 4 \quad E = 6 \quad F = 4$

$V + F = E + 2$

$4 + 4 = 6 + 2$

c. $V = 6 \quad E = 12 \quad F = 8$

$V + F = E + 2$

$6 + 8 = 12 + 2$

29. $V = 10 \cdot 3 \cdot 4 - 2(1 \cdot 1 \cdot 3)$

$V = 120 - 6$

$V = 114 \text{ in}^3$

30. a. $\frac{4}{8} = \frac{1}{2}$ b. $\frac{5}{8}$

31. a. $A = 6.5(12) = 78 \text{ in}^2$

b. $A = 4\left(\frac{s^2\sqrt{3}}{4}\right) = 4\left(\frac{4^2\sqrt{3}}{4}\right) = 16\sqrt{3} \text{ cm}^2$

Chapter Nine: Analytic Geometry

SECTION 9.1: The Rectangular Coordinate System

1\.

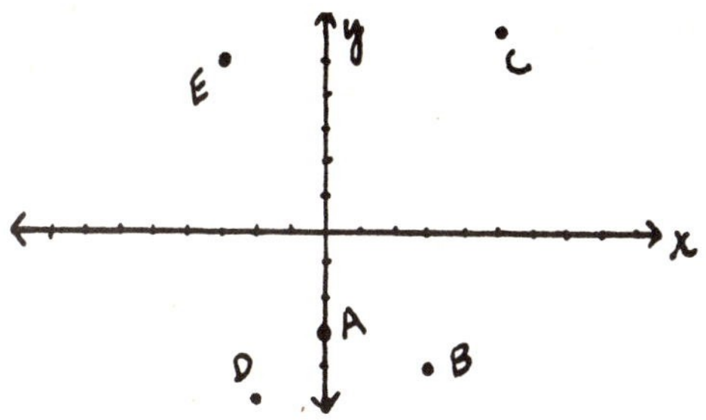

5\. The segment is horizontal so

$d = 7 - b$ if $7 > b$

$\therefore\ 7 - b = 3.5 \rightarrow -b = -3.5 \rightarrow$

$b = 3.5$. If $b > 7$, then $d = b - 7$.

$\therefore b - 7 = 3.5 \rightarrow b = 10.5$

9\. a. $M = \left(\frac{x_1 + x_2}{2}, \frac{y_1 + y_2}{2}\right)$

$M = \left(\frac{0 + 4}{2}, \frac{(-3) + 0}{2}\right)$

$M = \left(\frac{4}{2}, \frac{-3}{2}\right) = \left(2, \frac{-3}{2}\right)$

b. $M = \left(\frac{(-2) + 4}{2}, \frac{5 + (-3)}{2}\right)$

$M = \left(\frac{2}{2}, \frac{2}{2}\right)$

$M = (1,1)$

c. $M = \left(\frac{3 + 5}{2}, \frac{2 + (-2)}{2}\right)$

$M = \left(\frac{8}{2}, \frac{0}{2}\right)$

$M = (4,0)$

d. $M = \left(\frac{a + 0}{2}, \frac{0 + b}{2}\right)$

$M = \left(\frac{a}{2}, \frac{b}{2}\right)$

13\. a. $(4, -\frac{5}{2})$

b. $(0, 4)$

c. $(\frac{7}{2}, -1)$

d. (a, b)

17\. a. $(-3, -4)$

b. $(-2, 0)$

c. $(-a, 0)$

d. $(-b, c)$

21\. $M = \left(\frac{x_1 + x_2}{2}, \frac{y_1 + y_2}{2}\right)$

$(2.1, -5.7) = \left(\frac{x + 1.7}{2}, \frac{y + 2.3}{2}\right)$

$\frac{x + 1.7}{2} = 2.1$ and $\frac{y + 2.3}{2} = -5.7$

$x + 1.7 = 4.2$ and $y + 2.3 = -11.4$

$x = 2.5$ and $y = -13.7$

$B = (2.5, -13.7)$

25\. a. $AB = 4 - 0 = 4$

$$\begin{aligned} BC &= \sqrt{(2 - 4)^2 + (5 - 0)^2} \\ &= \sqrt{(-2)^2 + 5^2} \\ &= \sqrt{4 + 25} = \sqrt{29} \end{aligned}$$

$$\begin{aligned} AC &= \sqrt{(2 - 0)^2 + (5 - 0)^2} \\ &= \sqrt{2^2 + 5^2} \\ &= \sqrt{4 + 25} = \sqrt{29} \end{aligned}$$

Because $BC = AC$, $\triangle ABC$ is isosceles.

b. $DE = 4 - 0 = 4$

$DF = \sqrt{(2-0)^2 + (2\sqrt{3} - 0)^2}$

$= \sqrt{2^2 + (2\sqrt{3})^2}$

$= \sqrt{4 + 12} = \sqrt{16} = 4$

$EF = \sqrt{(2-4)^2 + (2\sqrt{3} - 0)^2}$

$= \sqrt{(-2)^2 + (2\sqrt{3})^2}$

$= \sqrt{4 + 12} = \sqrt{16} = 4$

Because $DE = DF = EF$, $\triangle DEF$ is equilateral.

c. $GH = \sqrt{(-2 - [-5])^2 + (6 - 2)^2}$

$= \sqrt{3^2 + 4^2}$

$= \sqrt{9 + 16} = \sqrt{25} = 5$

$GK = \sqrt{(2 - [-5])^2 + (3 - 2)^2}$

$= \sqrt{7^2 + 1^2} = \sqrt{50}$

$= \sqrt{25} \cdot \sqrt{2} = 5\sqrt{2}$

$HK = \sqrt{(2 - [-2])^2 + (3 - 6)^2}$

$= \sqrt{4^2 + (-3)^2} = \sqrt{16 + 9}$

$= \sqrt{25} = 5$

Because $GH = HK$ and $(GH)^2 + (HK)^2 = (GK)^2$, $\triangle GHK$ is an isosceles right triangle.

29. Call the third vertex (a,b). Because the three sides must be of the same length and $AB = 2a$, we have

$\sqrt{(a-0)^2 + (b-0)^2} = 2a$

$\sqrt{a^2 + b^2} = 2a$

Squaring, $a^2 + b^2 = 4a^2$

$b^2 = 3a^2$

$b = \pm\sqrt{3a^2}$

$b = \pm\sqrt{3} \cdot \sqrt{a^2}$

$b = \pm a\sqrt{3}$

The third vertex is $(a, a\sqrt{3})$ or $(a, -a\sqrt{3})$.

33. $A_{MNQ} = A_{RECT} - (A_{\triangle 1} + A_{\triangle 2} + A_{\triangle 3})$

$A = (7)(5) - [\frac{1}{2}(1)(5) + \frac{1}{2}(6)(4) + \frac{1}{2}(7)(1)]$

$A = 35 - [\frac{5}{2} + 12 + \frac{7}{2}]$

$A = 35 - 18$

$A = 17$

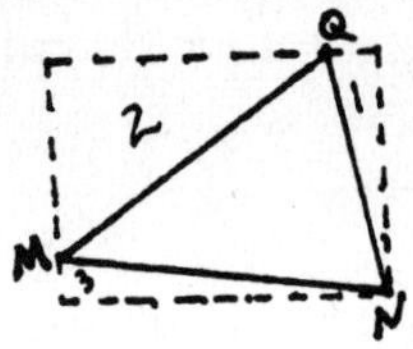

37. a. A cone is formed; $r = 9$; $h = 5$

$V = \frac{1}{3}\pi r^2 h$

$V = \frac{1}{3}\pi \cdot 9^2 \cdot 5$

$V = 135\pi$ units^3

b. A cone is formed; $r = 5$; $h = 9$

$V = \frac{1}{3}\pi r^2 h$

$V = \frac{1}{3}\pi \cdot 5^2 \cdot 9$

$V = 75\pi$ units^3

41. a. $L = 2\pi rh$

$L = 2\pi \cdot 5 \cdot 9$

$L = 90\pi$ units^2

b. $L = 2\pi rh$

$L = 2\pi \cdot 9 \cdot 5$

$L = 90\pi$ units^2

SECTION 9.2: Graphs of Linear Equations and Slope

1. $3x + 4y = 12$ has intercepts (4,0) and (0,3).

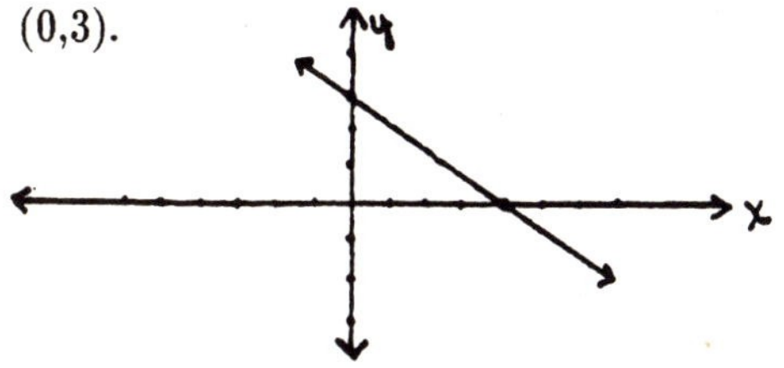

5. $2x + 6 = 0$ is equivalent to $x = -3$. It is a vertical line with x-intercept $(-3,0)$.

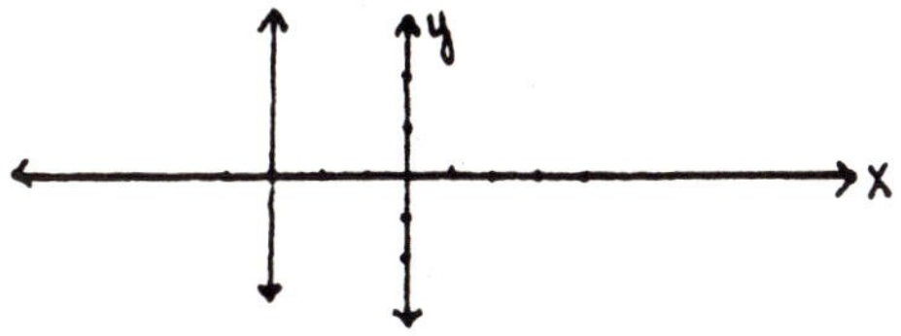

9. a. $m = \frac{y_2 - y_1}{x_2 - y_1}$

$m = \frac{5 - (-3)}{4 - 2}$

$m = \frac{8}{2} \rightarrow m = 4$

b. $m = \frac{7 - (-2)}{3 - 3} = \frac{9}{0}$

m is undefined

c. $m = \frac{-2 - (-1)}{2 - 1}$

$m = \frac{-1}{1} \rightarrow m = -1$

d. $m = \frac{5 - 5}{(-1.3) - (-2.7)}$

$m = \frac{0}{1.4} \rightarrow m = 0$

e. $m = \frac{d - b}{c - a}$

f. $m = \frac{b - 0}{0 - a} = \frac{b}{-a} \rightarrow m = -\frac{b}{a}$

13. a. $m_{\overline{AB}} = \frac{2 - 5}{0 - (-2)}$

$m_{\overline{AB}} = -\frac{3}{2}$

$m_{\overline{BC}} = \frac{-4 - 2}{4 - 0}$

$m_{\overline{BC}} = \frac{-6}{4}$

$m_{\overline{BC}} = -\frac{3}{2}$

Because $m_{\overline{AB}} = m_{\overline{BC}}$, the points A, B, and C are collinear.

b. $m_{\overline{DE}} = \frac{-2 - (-1)}{2 - (-1)}$

$m_{\overline{DE}} = \frac{-1}{3}$ or $-\frac{1}{3}$

$m_{\overline{EF}} = \frac{(-5) - (-2)}{5 - 2}$

$m_{\overline{EF}} = \frac{-3}{3} = -1$

Because $m_{\overline{DE}} \neq m_{\overline{EF}}$, the points D, E, and F are noncollinear.

17. $2x + 3y = 6$ contains the points

(3,0) and (0,2) so that its slope is

$m_1 = \frac{2 - 0}{0 - 3} = -\frac{2}{3}$

$2x - 3y = 12$ contains (6,0) and $(0, -4)$, so that its slope is

$m_2 = \frac{-4 - 0}{0 - 6} = \frac{-4}{-6} = \frac{2}{3}$

None of these

21. Points A, B, and C are collinear if

$m_{\overline{AB}} = m_{\overline{BC}}$.

$m_{\overline{AB}} = \frac{3-5}{2-x} = \frac{-2}{2-x}$.

$m_{\overline{BC}} = \frac{-5-3}{4-2} = \frac{-8}{2} = \frac{-4}{1}$ or -4

$\frac{-2}{2-x} = 4$

Multiply by $2 - x$

$-2 = -4(2 - x)$

$-2 = -8 + 4x$

$4x = 6$

$x = \frac{3}{2}$

25. First plot the point $(3, -2)$. If $m = 2$ or $m = \frac{2}{1}$, then a change in y of 2 corresponds to a change in x of 1.

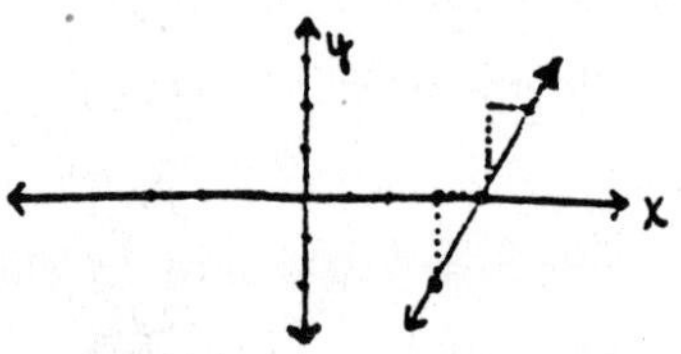

29. The line $2x - y = 6$ contains (3,0) and $(0, -6)$, so its slope is
$m = \frac{-6-0}{0-3} = \frac{-6}{-3} = 2$.
Plot the point $(-2,1)$. Then draw the line for which a change of 2 in y corresponds to a change of 1 in x.

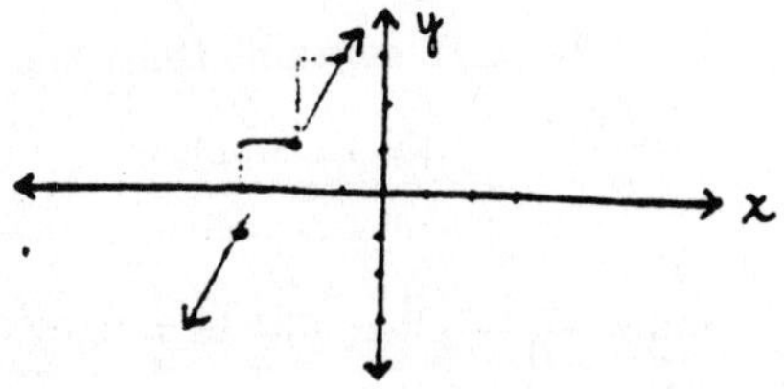

33. As shown, possible locations are $P_1 = (4,7)$
$P_2 = (0, -1)$, and
$P_3 = (10, -3)$.

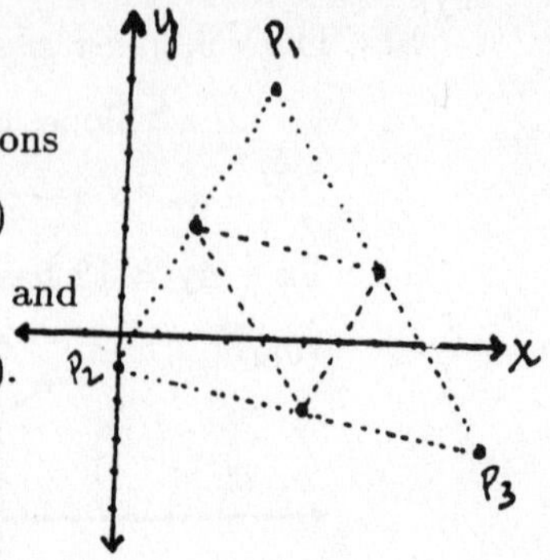

37. $m_{\overline{EH}} = \frac{2c - 0}{2b - 0} = \frac{2c}{2b} = \frac{c}{b}$

$m_{\overline{FG}} = \frac{c - 0}{(a + b) - a} = \frac{c}{b}$

Since one pair of opposite sides are parallel, EFGH is a trapezoid.

SECTION 9.3: Preparing to do Analytic Proofs

1. a. $d = \sqrt{(x_2 - x_1)^2 + (y_2 - y_1)^2}$

$d = \sqrt{(0 - a)^2 + (a - 0)^2}$

$d = \sqrt{(-a)^2 + a^2}$

$d = \sqrt{a^2 + a^2}$

$d = \sqrt{2a^2} = a\sqrt{2}$ if $a > 0$

b. $m = \frac{y_2 - y_1}{x_2 - x_1}$

$m = \frac{d - b}{c - a}$

5. $\overline{AB}$ is horizontal and $\overline{BC}$ is vertical $\therefore \overline{AB} \perp \overline{BC}$. Hence, $\angle B$ is a right $\angle$ and $\triangle ABC$ is a right triangle.

9 . $m_{\overline{MN}} = 0$ and $m_{\overline{QP}} = 0$ $\therefore$ $\overline{MN} \parallel \overline{QP}$

$\overline{QM}$ and $\overline{PN}$ are both vertical;

$\therefore$ $\overline{QM} \parallel \overline{PN}$. Hence, MQPN is a parallelogram. Since $\overline{QM}$ is vertical and $\overline{MN}$ is horizontal, $\angle$QMN is a right angle. Because parallelogram MQPN has a right $\angle$, it is also a rectangle.

13. M = (0, 0); N = (r, 0); P = (r + s, t)

17. a. Square

A = (0, 0)

B = (a, 0)

C = (a, a)

D = (0, a)

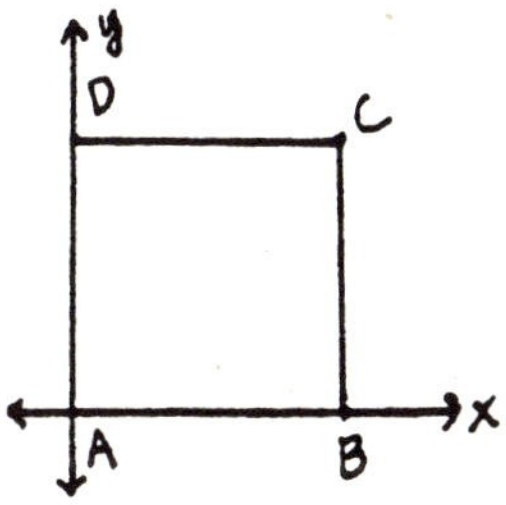

b. Square (with midpoints of sides)

A = (0, 0); B = (2a, 0)

C = (2a, 2a); D = (0, 2a)

21. a. Isosceles triangle

R = (0, 0)

S = (2a, 0)

T = (a, b)

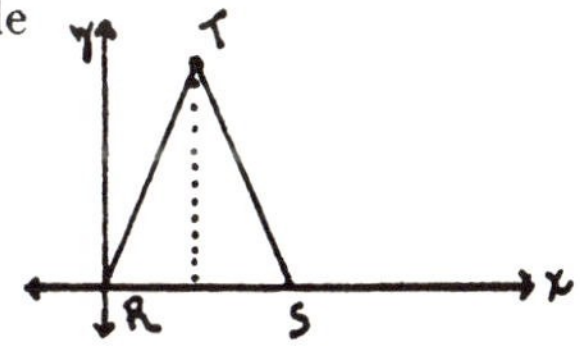

b. Isosceles triangle (with midpoints)

R = (0, 0)

S = (4a, 0)

T = (2a, 2b)

25. Parallelogram ABCD with $\overline{AC} \perp \overline{DB}$.

$$m_{\overline{AC}} = \frac{c - 0}{a + b - 0} = \frac{c}{a + b}$$

$$m_{\overline{DB}} = \frac{0 - c}{a - b} = \frac{-c}{a - b}$$

$$\therefore \frac{c}{a + b} \cdot \frac{-c}{a - b} = -1$$

or

$$\frac{-c^2}{a^2 - b^2} = -1$$

$$-c^2 = -(a^2 - b^2)$$

$$c^2 = a^2 - b^2$$

29. a. a is positive

b. $-a$ is negative

c. AB = a − (−a) = 2a

33. The line segment joining the midpoints of the two nonparallel sides of a trapezoid is parallel to the bases of the trapezoid.

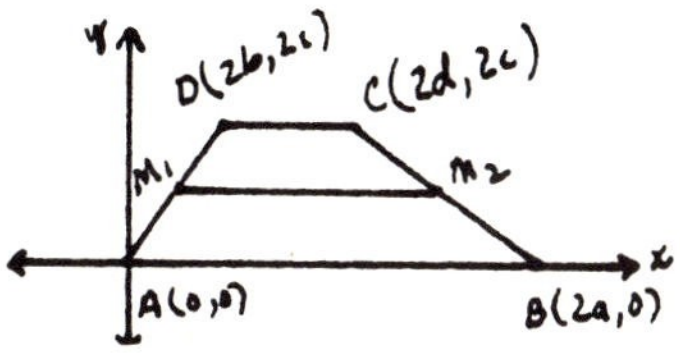

SECTION 9.4: Analytic Proofs

1 . The diagonals of a rectangle are equal in length.

Proof: Let rectangle ABCD have vertices as shown.

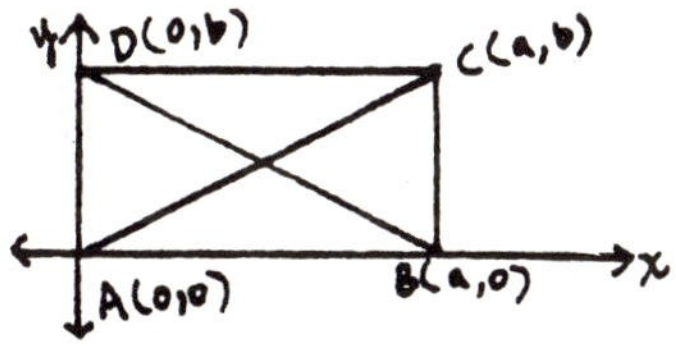

Then $AC = \sqrt{(a - 0)^2 + (b - 0)^2}$
$= \sqrt{a^2 + b^2}$

Also $DB = \sqrt{(a - 0)^2 + (0 - b)^2}$
$\sqrt{a^2 + (-b^2)} = \sqrt{a^2 + b^2}$

Then AC = DB, and the diagonals of the rectangle are equal in length.

5. The median from the vertex of an isosceles triangle to the base is perpendicular to the base.

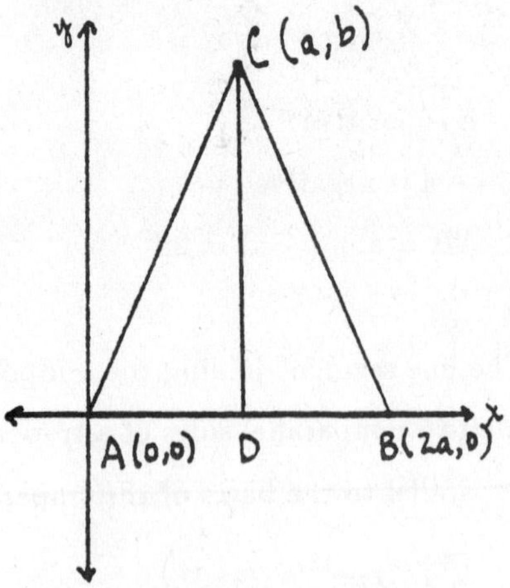

Proof: The triangle ABC with vertices as shown is isosceles because AC = BC.

Let D be the midpoint of $\overline{AB}$.

$$D = \left(\frac{0 + 2a}{2}, \frac{0 + 0}{2}\right) = (a,0).$$

$m_{\overline{AB}} = \frac{0 - 0}{2a - 0} = 0$; that is $\overline{AB}$ is horizontal.

$m_{\overline{CD}} = \frac{b - 0}{a - a} = \frac{b}{0}$ which is undefined; that is $\overline{CD}$ is vertical.

Then $\overline{CD} \perp \overline{AB}$ and the median to the base of the isosceles triangle is perpendicular to the base.

9. The segments which join the midpoints of the consecutive sides of a rectangle form a rhombus.

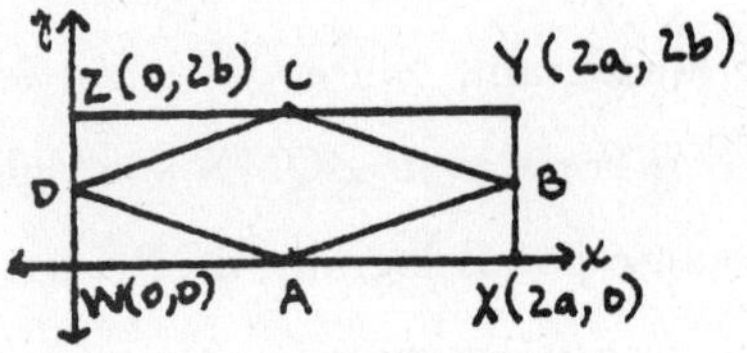

Proof: With vertices as shown, WXYZ is a rectangle.

Midpoints of the sides of the rect. are

$$A = \left(\frac{0 + 2a}{2}, \frac{0 + 0}{2}\right) = (a,0),$$

$$B = \left(\frac{2a + 2a}{2}, \frac{0 + 2b}{2}\right) = (2a,b),$$

$$C = \left(\frac{0 + 2a}{2}, \frac{2b + 2b}{2}\right) = (a,2b), \text{ and}$$

$$D = \left(\frac{0 + 0}{2}, \frac{0 + 2b}{2}\right) = (0,b).$$

$m_{\overline{AB}} = \frac{b - 0}{2a - a} = \frac{b}{a}$ and

$m_{\overline{DC}} = \frac{b - a}{0 - a} = \frac{-b}{-a} = \frac{b}{a}$, so $\overline{DA} \parallel \overline{CB}$.

$m_{\overline{DA}} = \frac{b - 0}{0 - a} = \frac{b}{a}$ and

$m_{\overline{CB}} = \frac{2b - b}{a - 2a} = \frac{b}{-a}$, so $\overline{DA} \parallel \overline{CB}$.

Then ABCD is a parallelogram.

We need to show that 2 adjacent sides are congruent (equal in length).

$$AB = \sqrt{(2a - a)^2 + (b - 0)^2} = \sqrt{a^2 + b^2}$$

$$BC = \sqrt{(a - 2a)^2 + (2b - b)^2} = \sqrt{(-a)^2 + b^2} = \sqrt{a^2 + b^2}$$

Then $\overline{AB} \doteq \overline{BC}$ and ABCD is a rhombus.

13. The segment which joins the midpoints of 2 sides of a triangle is parallel to the third side and has a length equal to one-half the length of the third side.

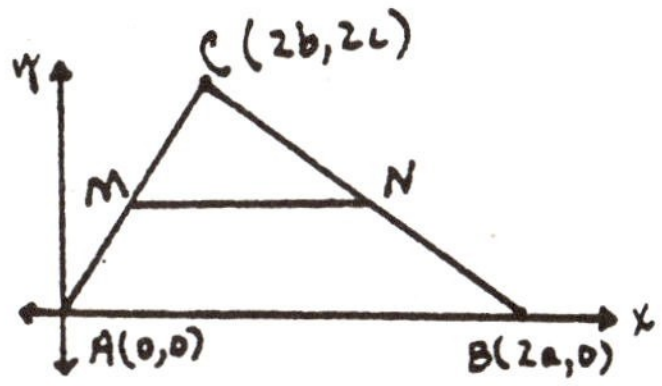

Proof: Let $\triangle ABC$ have vertices as shown. With M and N the midpoints of $\overline{AC}$ and $\overline{BC}$ respectively,

$$M = \left(\frac{0+2b}{2}, \frac{0+2c}{2}\right) = (b,c) \text{ and}$$

$$N = \left(\frac{2a+2b}{2}, \frac{0+2c}{2}\right) = (a+b,c).$$

Now $m_{\overline{MN}} = \frac{c-c}{(a+b)-b} = \frac{0}{a} = 0$ and

$$m_{\overline{AB}} = \frac{0-0}{2a-0} = \frac{0}{2a} = 0.$$

Then $\overline{MN} \parallel \overline{AB}$.

Also $MN = (a+b) - b = a$.

$AB = 2a - 0 = 2a$.

Then $MN = \frac{1}{2}(AB)$

That is, the segment $(\overline{MN})$ which joins the midpoints of 2 sides of the triangle is parallel to the third side and equals one-half its length.

17. If the diagonals of a parallelogram are perpendicular, then the parallelogram is a rhombus.

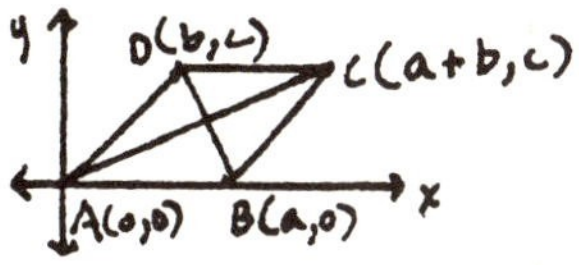

Proof: Let parallelogram ABCD have vertices as shown.

If $\overline{AC} \perp \overline{DB}$, then

$m_{\overline{AC}} \cdot m_{\overline{DB}} = -1.$

Because $m_{\overline{AC}} = \frac{c-0}{(a+b)-0} = \frac{c}{a+b}$

and

$$m_{\overline{DB}} = \frac{0-c}{a-b} = \frac{-c}{a-b},$$

it follows that

$$\frac{c}{a+b} \cdot \frac{-c}{a-b} = -1$$

or $\frac{-c^2}{a^2-b^2} = -1$. Then

$-c^2 = -1(a^2-b^2)$ or $-c^2 = -a^2 + b^2$

or $a^2 = b^2 + c^2$ (*).

For ABCD to be a rhombus, we must show that two adjacent sides are congruent.

$AB = a - 0 = a$

$AD = \sqrt{(b-0)^2 + (c-0)^2} = \sqrt{b^2+c^2}$

Because $a^2 = b^2 + c^2(*)$, it follows that $a = \sqrt{b^2+c^2}$ and $AB = AD$.

Then parallelogram ABCD is a rhombus.

21. The theorem remains true.
The quadrilateral which results is a parallelogram.

SECTION 9.5: Equations of Lines

1. Dividing by 8,

$8x + 16y = 48$

becomes

$x + 2y = 6$

Then $2y = -1x + 6$

$\frac{1}{2}(2y) = \frac{1}{2}(-1x + 6)$

$y = -\frac{1}{2}x + 3$

5. $y = 2x - 3$

has $m = 2$ and $b = -3$. Plot the point $(0, -3)$. Then draw the line for which an increase of 2 in y corresponds to an increase of 1 in x.

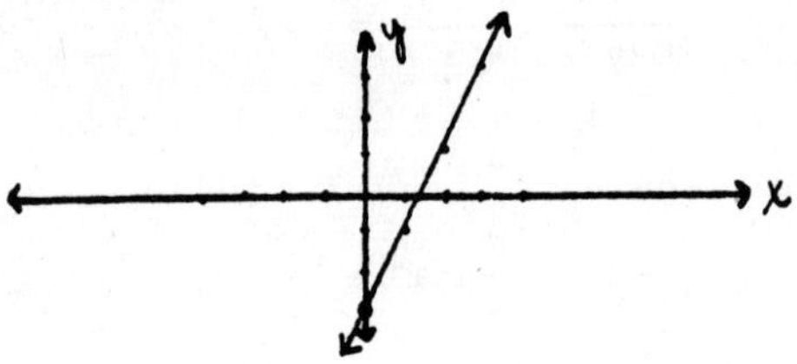

9. Using $y = mx + b$,

$y = -\frac{2}{3}x + 5.$

$3y = -2x + 15$

$2x + 3y = 15$

13. $m = \frac{1 - (-1)}{3 - 0} = \frac{2}{3}$

Using $y = mx + b$,

$y = \frac{2}{3}x - 1$

Mult. by 3,

$3y = 2x - 3$

$-2x + 3y = -3$

17. The graph contains (2,0) and (0, −2).

$m = \frac{-2 - 0}{0 - 2} = \frac{-2}{-2} = 1.$

Using $y = mx + b$,

$y = 1x - 2$

$-x + y = -2$

21. The line $y = \frac{3}{4}x - 5$ has slope $m_1 = \frac{3}{4}$

The desired line has $m_2 = -\frac{4}{3}$.

Using $y = mx + b$, $y = -\frac{4}{3}x - 4$

Mult. by 3, $3y = -4x - 12$

$4x + 3y = -12$

25. Perpendicular slope is $-\frac{b}{a}$.

$y - h = -\frac{b}{a}(x - g)$

$y - h = -\frac{b}{a}x + \frac{bg}{a}$

$y = -\frac{b}{a}x + \frac{bg + ha}{a}$

29. $2x + y = 6 \rightarrow y = -2x + 6$

$3x - y = 19 \rightarrow y = 3x - 19$

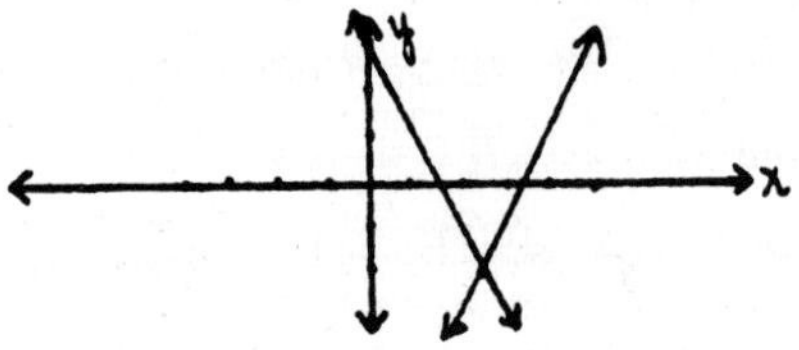

The lines intersect at (5, −4).

33. $2x + y = 8$

$\underline{3x - y = 7}$

$5x = 15$

$x = 3$

$2(3) + y = 8$

$6 + y = 8 \rightarrow y = 2$

The point of intersection is (3,2).

37. $2x + 3y = 4$ Multiply by -3

$3x - 4y = 23$ Multiply by 2

$-6x - 9y = -12$

$6x - 8y = 46$

$-17y = 34$

$y = -2$

$2x + 3(-2) = 4$

$2x - 6 = 4$

$2x = 10$

$x = 5$

The point of intersection is $(5, -2)$.

41. $bx + c = a$

$bx = a - c$

$x = \frac{a - c}{b}$ and $y = a$

$\left(\frac{a - c}{b}, a\right)$

45. The altitudes of a triangle are concurrent.

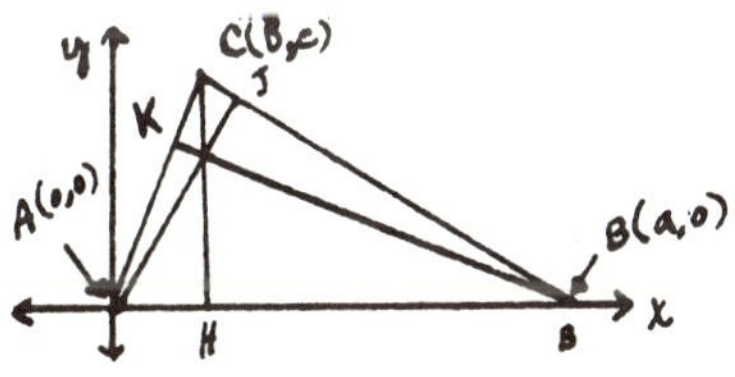

Proof: For $\triangle ABC$, let $\overline{CH}$, $\overline{AJ}$, and $\overline{BK}$ name the altitudes. Because $\overline{AB}$ is horizontal ($m_{\overline{AB}} = 0$), $\overline{CH}$ is vertical and has the equation $x = b$. Because $m_{\overline{BC}} = \frac{c - 0}{b - a} = \frac{c}{b - a}$, the slope of altitude $\overline{AJ}$ is $m_{\overline{AJ}} = -\frac{b - a}{c} = \frac{a - b}{c}$.

Since $\overline{AJ}$ contains (0,0), its equation is

$y = \frac{a - b}{c}x$.

The intersection of altitudes $\overline{CH}$ ($x = b$) and $\overline{AJ}$ ($y = \frac{a - b}{c}x$) is at $x = b$ so that

$y = \frac{a - b}{c} \cdot b = \frac{b(a - b)}{c} = \frac{ab - b^2}{c}$.

That is, $\overline{CH}$ and $\overline{AJ}$ intersect at $\left(b, \frac{ab - b^2}{c}\right)$. The remaining altitude is $\overline{BK}$. Since $m_{\overline{AC}} = \frac{c - 0}{b - 0} = \frac{c}{b}$, $m_{\overline{BK}} = -\frac{b}{c}$. Because $\overline{BK}$ contains (a,0), its equation is $y - 0 = -\frac{b}{c}(x - a)$ or

$y = \frac{-b}{c}(x - a)$.

For the three altitudes to be concurrent, $\left(b, \frac{ab - b^2}{c}\right)$ must lie on the line $y = \frac{-b}{c}(x - a)$. Substitution leads to

$\frac{ab - b^2}{c} = \frac{-b}{c}(b - a)$

$\frac{ab - b^2}{c} = \frac{-b(b - a)}{c}$

$\frac{ab - b^2}{c} = \frac{-b^2 + ab}{c}$, which is true.

Therefore the three altitudes are concurrent.

1. a. $d = y_2 - y_1$ (if $y_2 > y_1$) $= 4 - (-3) = 7$

   b. $d = x_2 - x_1$ (if $x_2 > x_1$) $= 1 - (-5) = 6$

   c. $d = \sqrt{(x_2 - x_1)^2 + (y_2 - y_1)^2}$

   $d = \sqrt{(7 - [-5])^2 + (-3 - 2)^2}$

   $d = \sqrt{12^2 + (-5)^2} = \sqrt{144 + 25} = \sqrt{169} = 13$

   d. $d = \sqrt{[(x - (x - 3)]^2 + [(y - 2) - (y + 2)]^2}$

   $d = \sqrt{(3)^2 + (-4)^2} = \sqrt{9 + 16} = \sqrt{25} = 5$

2. a. $d = y_2 - y_1$ (if $y_2 > y_1$) $= 5 - (-3) = 8$

   b. $d = x_2 - x_1$ (if $x_2 > x_1$) $= 3 - (-7) = 10$

   c. $d = \sqrt{[4 - (-4)]^2 + [5 - 1]^2} = \sqrt{8^2 + 4^2} = \sqrt{64 + 16} = \sqrt{80} = \sqrt{16} \cdot \sqrt{5} = 4\sqrt{5}$

   d. $d = \sqrt{[(x + 4) - (x - 2)]^2} + \sqrt{[(y + 5) - (y - 3)]^2} = \sqrt{6^2 + 8^2} = \sqrt{36 + 64} = \sqrt{100} = 10$

3. a. $M = \left(\frac{x_1 + x_2}{2}, \frac{y_1 + y_2}{2}\right)$

   $M = \left(\frac{6 + 6}{2}, \frac{4 + (-3)}{2}\right) = \left(6, \frac{1}{2}\right)$

   b. $M = \left(\frac{1 + (-5)}{2}, \frac{4 + 4}{2}\right) = (-2, 4)$

   c. $M = \left(\frac{(-5) + 7}{2}, \frac{2 + (-3)}{2}\right) = \left(1, \frac{-1}{2}\right)$

   d. $M = \left(\frac{(x - 3) + x}{2}, \frac{(y + 2) + (y - 2)}{2}\right) = \left(\frac{2x - 3}{2}, y\right)$

4. a. $M = \left(\frac{2+2}{2}, \frac{(-3)+5}{2}\right) = (2, 1)$

b. $M = \left(\frac{3+(-7)}{2}, \frac{(-2)+(-2)}{2}\right) = (-2, -2)$

c. $M = \left(\frac{(-4)+4}{2}, \frac{1+5}{2}\right) = (0, 3)$

d. $M = \left(\frac{(x-2)+(x+4)}{2}, \frac{(y-3)+(y+5)}{2}\right)$

$= (x+1, y+1)$

5. a. $m = \frac{y_2 - y_1}{x_2 - x_1}$

$m = \frac{(-3)-4}{6-6} = \frac{-7}{0}$ m is undefined

b. $m = \frac{4-4}{-5-1} = \frac{0}{-6} = 0$

c. $m = \frac{-3-2}{7-(-5)} = \frac{-5}{12}$

d. $m = \frac{(y-2)-(y+2)}{x-(x-3)} = \frac{-4}{3}$

6. a. $m = \frac{5-(-3)}{2-2} = \frac{8}{0}$ m is undefined

b. $m = \frac{(-2)-(-2)}{-7-3} = \frac{0}{-10} = 0$

c. $m = \frac{5-1}{4-(-4)} = \frac{4}{8} = \frac{1}{2}$

d. $m = \frac{(y+5)-(y-3)}{(x+4)-(x-2)} = \frac{8}{6} = \frac{4}{3}$

7. $M = \left(\frac{x_1+x_2}{2}, \frac{y_1+y_2}{2}\right)$

$(2,1) = \left(\frac{8+x}{2}, \frac{10+y}{2}\right)$

$\frac{8+x}{2} = 2$ and $\frac{10+y}{2} = 1$

$8 + x = 4$ and $10 + y = 2$

$x = -4$ and $y = -8$

$B = (-4, -8)$

8. Due to symmetry, R = (3, 7).

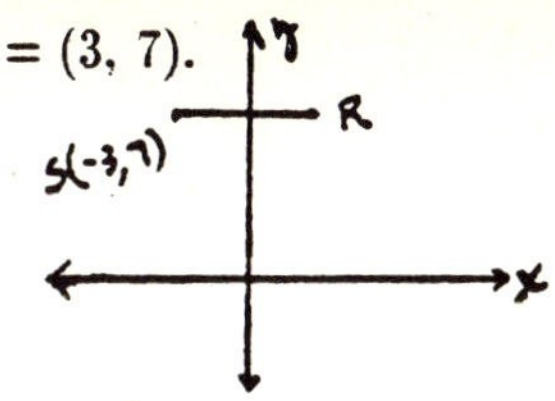

9. $m = \frac{y_2 - y_1}{x_2 - x_1}$

$-3 = \frac{3-1}{x-2}$ or $-3 = \frac{2}{x-2}$

$-3(x-2) = 2$

$-3x + 6 = 2$

$-3x = -4$

$x = \frac{4}{3}$

10. $m = \frac{y_2 - y_1}{x_2 - x_1}$

$\frac{-6}{7} = \frac{y-2}{2-(-5)}$

$\frac{-6}{7} = \frac{y-2}{7}$

$-6 = y - 2$

$y = -4$

11. a. $x + 3y = 6 \rightarrow y = -\frac{1}{3}x + 2; m = -\frac{1}{3}$

$3x - y = -7 \rightarrow y = 3x + 7; m = 3$

Since $m_1 \cdot m_2 = -1$, the lines are perpendicular.

b. $2x - y = -3 \rightarrow y = 2x + 3 \rightarrow$

$m = 2$ and $b = 3$

$y = 2x - 14 \rightarrow m = 2$ and $b = -14$.

Since $m_1 = m_2$, the lines are parallel.

c. $y + 2 = -3(x-5) \rightarrow y = -3x + 13$

$\rightarrow m = -3$

$2y = 6x + 11 \rightarrow y = 3x + \frac{11}{2} \rightarrow m = 3$

The lines are neither parallel nor perpendicular.

d. $.5x + y = 0 \rightarrow y = \frac{-1}{2}x \rightarrow m = \frac{-1}{2}$

$2x - y = 10 \rightarrow y = 2x - 10 \rightarrow m = 2$

Since $m_1 \cdot m_2 = -1$, the lines are $\perp$.

12. Let A = (− 6, 5), B = (1, 7), and C = (16, 10). The points would be collinear if

$m_{\overline{AB}} = m_{\overline{BC}}$.

$$m_{\overline{AB}} = \frac{7 - 5}{1 - (-6)} = \frac{2}{7};$$

$$m_{\overline{BC}} = \frac{10 - 7}{16 - 1} = \frac{3}{15} = \frac{1}{5}.$$

The points are not collinear.

13. Let A = (− 2, 3), B = (x, 6), and C = (8, 8).

If $m_{\overline{AB}} = m_{\overline{BC}}$, then A, B, and C are collinear.

$$m_{\overline{AB}} = \frac{6 - 3}{x - (-2)} = \frac{3}{x + 2};$$

$$m_{\overline{BC}} = \frac{8 - 6}{8 - x} = \frac{2}{8 - x}$$

$$\frac{3}{x + 2} = \frac{2}{8 - x}$$

$$3(8 - x) = 2(x + 2)$$

$$24 - 3x = 2x + 4$$

$$20 = 5x$$

$$x = 4$$

14. Intercepts are (7, 0) and (0, 3).

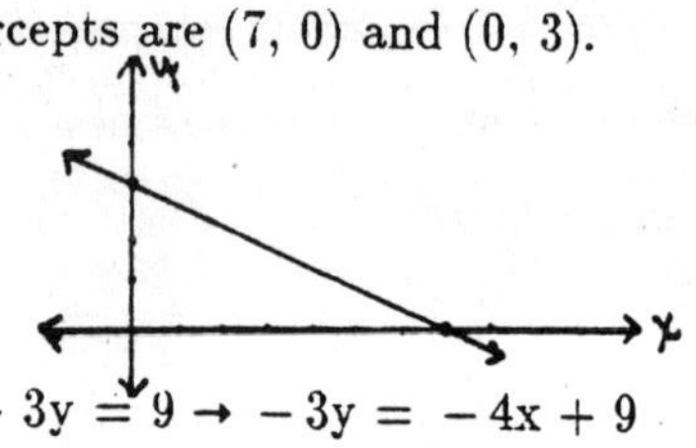

15. $4x - 3y = 9 \rightarrow -3y = -4x + 9$

$$\rightarrow y = \frac{4}{3}x - 3$$

First, plot the point (0, − 3). Then locate a second point for which an increase of 4 in y corresponds to an increase of 3 in x.

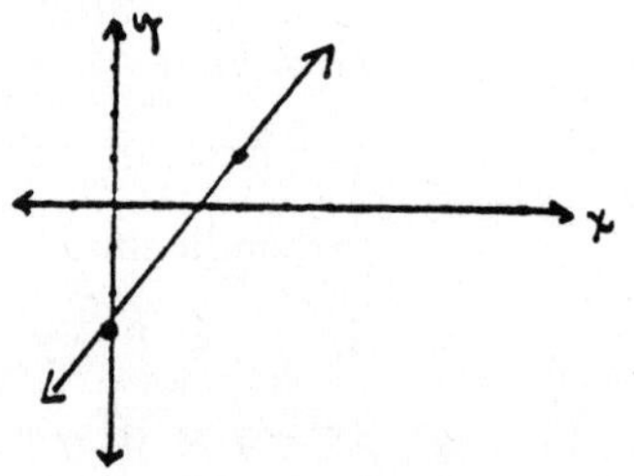

16. $y + 2 = \frac{-2}{3}(x - 1)$

$y - (-2) = \frac{-2}{3}(x - 1)$ is a line which contains (1, − 2) and has slope $m = \frac{-2}{3}$.

1st Plot (1, − 2). Then from that point draw a line which has $m = \frac{-2}{3}$.

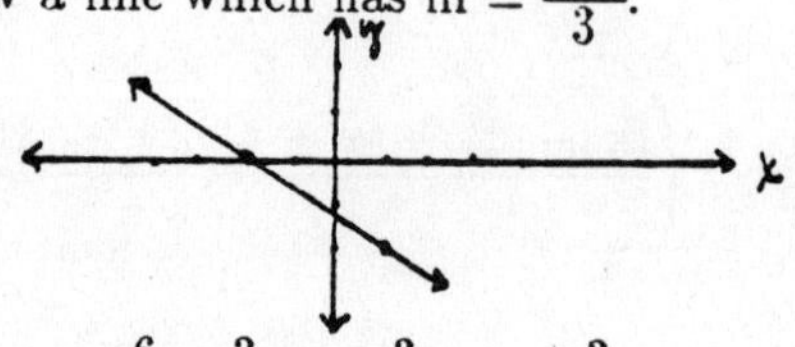

17. a. $m = \frac{6 - 3}{-3 - 2} = \frac{3}{-5}$ or $\frac{-3}{5}$.

$$y - 3 = \frac{-3}{5}(x - 2)$$

$$5(y - 3) = -3(x - 2)$$

$$5y - 15 = -3x + 6$$

$$3x + 5y = 21$$

b. $m = \frac{(-9) - (-3)}{8 - 6} = \frac{-6}{2} = -3$

$$y - [-1] = -3(x - [-2])$$

$$y + 1 = -3(x + 2)$$

$$y + 1 = -3x - 6$$

$$3x + y = -7$$

c. $x + 2y = 4 \rightarrow y = \frac{-1}{2}x + 2$

Since $m_1 = \frac{-1}{2}$, the desired line has $m_2 = 2$.

$$y - [-2] = 2(x - 3)$$

$$y + 2 = 2x - 6$$

$$-2x + y = -8$$

d. A line parallel to the x-axis has the form y = b. ∴ y = 5

18. Let A = $(-2, -3)$, B = $(4, 5)$, and C = $(-4, 1)$.

$$m_{\overline{AB}} = \frac{5-(-3)}{4-(-2)} = \frac{8}{6} = \frac{4}{3};$$

$$m_{\overline{BC}} = \frac{1-5}{-4-4} = \frac{-4}{-8} = \frac{1}{2}; \text{ and}$$

$$m_{\overline{AC}} = \frac{1-(-3)}{(-4)-(-2)} = \frac{4}{-2} = -2.$$

Because $m_{\overline{AC}} \cdot m_{\overline{BC}} = -1$, $\overline{AC} \perp \overline{BC}$ and $\angle C$ is a rt. $\angle$.

19. Let A = $(3, 6)$, B = $(-6, 4)$, and C = $(1, -2)$.

$$AB = \sqrt{(-6-3)^2 + (4-6)^2} = \sqrt{(-9)^2 + (-2)^2} = \sqrt{81+4} = \sqrt{85}$$

$$BC = \sqrt{[1-(-6)]^2 + [(-2)-4]^2} = \sqrt{7^2 + (-6)^2} = \sqrt{49+36} = \sqrt{85}$$

$$AC = \sqrt{(1-3)^2 + ([-2]-6)^2} = \sqrt{(-2)^2 + (-8)^2} = \sqrt{4+64} = \sqrt{68} = \sqrt{4 \cdot 17} = 2\sqrt{17}$$

Because AB = BC, the triangle is isosceles.

20. R = $(-5, -3)$, S = $(1, -11)$, T = $(7, -6)$ and V = $(1, 2)$.

$$m_{\overline{RS}} = \frac{(-11)-(-3)}{1-(-5)} = \frac{-8}{6} = \frac{-4}{3};$$

$$m_{\overline{ST}} = \frac{(-6)-(-11)}{7-1} = \frac{5}{6};$$

$$m_{\overline{TV}} = \frac{2-(-6)}{1-7} = \frac{8}{-6} = \frac{-4}{3};$$

$$m_{\overline{RV}} = \frac{2-(-3)}{1-(-5)} = \frac{5}{6}$$

$\therefore$ $\overline{RS} \parallel \overline{VT}$ and $\overline{RV} \parallel \overline{ST}$ and RSTV is a parallelogram.

21. Solution by graphing:

$4x - 3y = -3 \rightarrow y = \frac{4}{3}x + 1$

$x + 2y = 13 \rightarrow y = \frac{-1}{2}x + \frac{13}{2}$

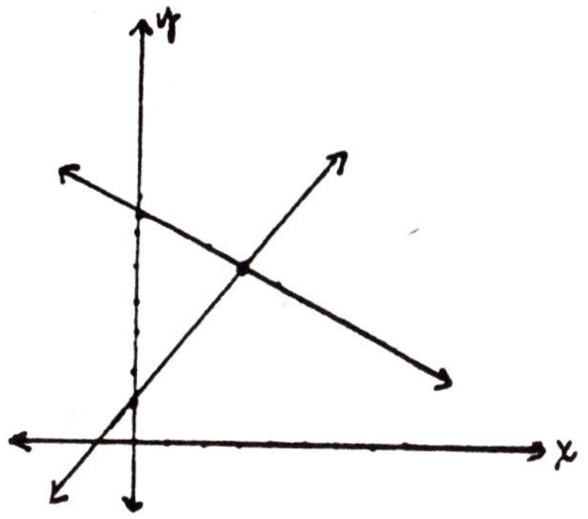

The graphs (lines) intersect at (3, 5).

22. Solution by graphing:

$y = x + 3$

$y = 4x$

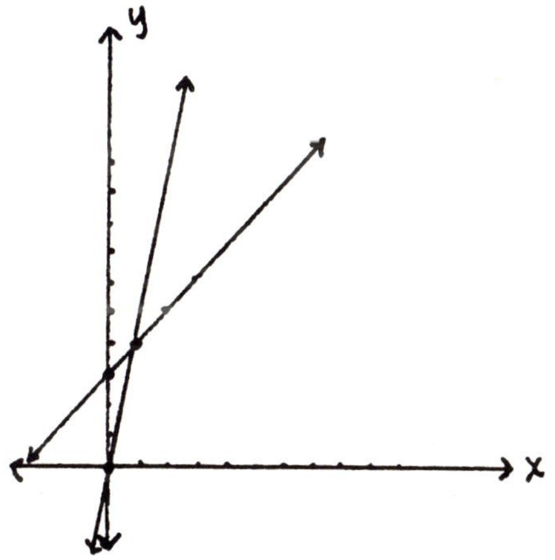

The graphs (lines) intersect at (1, 4).

23. Solution by Substitution:

$4x - 3y = -3$

$x + 2y = 13 \rightarrow x = 13 - 2y$

$4(13 - 2y) - 3y = -3$

$52 - 8y - 3y = -3$

$-11y = -55$

$y = 5$

$x = 13 - 2(5) = 3$

(3, 5)

24. Solution by Substitution:

$4x = x + 3$

$3x = 3$

$x = 1$

$y = 4(1) = 4$

(1, 4)

25.

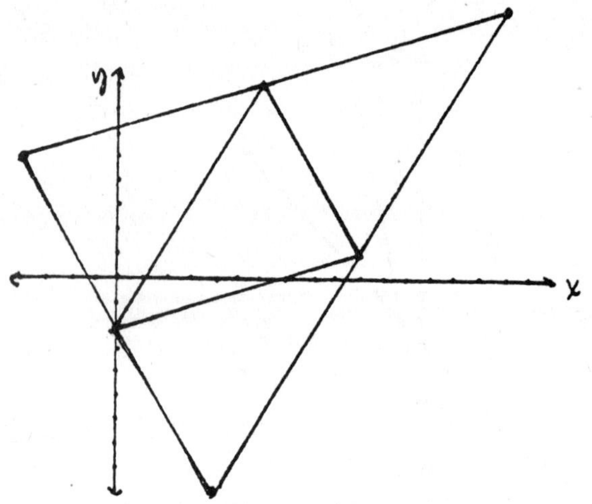

Possible pts. are (16,11), (4, −9), (−4,5).

26. a. $D = M_{\overline{AC}} = (7, 2)$

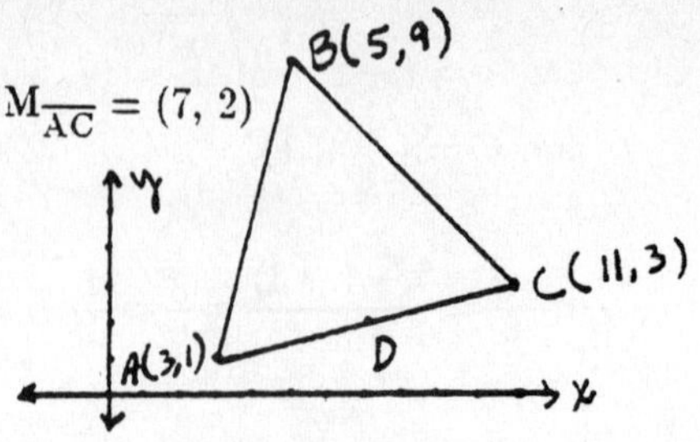

The length of $\overline{BD}$ is
$\sqrt{(7 - 5)^2 + (2 - 9)^2} =$
$\sqrt{2^2 + (-7)^2} = \sqrt{4 + 49} = \sqrt{53}.$

b. $m_{\overline{AC}} = \frac{3 - 1}{11 - 3} = \frac{2}{8} = \frac{1}{4}.$
Then the slope of the altitude to $\overline{AC}$ is -4.

c. Since $m_{\overline{AC}} = \frac{1}{4}$, the slope of any line parallel to $\overline{AC}$ is also $\frac{1}{4}$.

27. $A = (-a, 0)$
$B = (0, b)$
$C = (a, 0)$

28. $D = (0, 0)$
$E = (a, 0)$
$F = (a, 2a)$
$G = (0, 2a)$

29. $R = (0, 0)$
$U = (0, a)$
$T = (a, a + b)$

30. $M = (0, 0)$
$N = (a, 0)$
$Q = (a + b, c)$
$P = (b, c)$

31. a. The midpoint of $\overline{AB}$ is

$M_{\overline{AB}} = (a + c, b + d)$. Then CM =

$\sqrt{[(a + c) - 0]^2 + [(b + d) - 2e]^2}$

$= \sqrt{(a + c)^2 + (b + d - 2e)^2}$

b. $m_{\overline{AC}} = \frac{2e - 2b}{0 - 2a}$

$= \frac{e - b}{-a}$ or $\frac{b - e}{a}$

Then the slope of the altitude to $\overline{AC}$ is $-\frac{a}{b - e}$ or $\frac{a}{e - b}$.

c. The altitude from B to $\overline{AC}$ contains

the point (2c,2d) and has slope

$m = \frac{a}{e - b}$ from part (b).

$y - 2d = \frac{a}{e - b}(x - 2c)$.

32. See Section 9.4, #18.

33. If the diagonals of a rectangle are perpendicular, then the rectangle is a square.

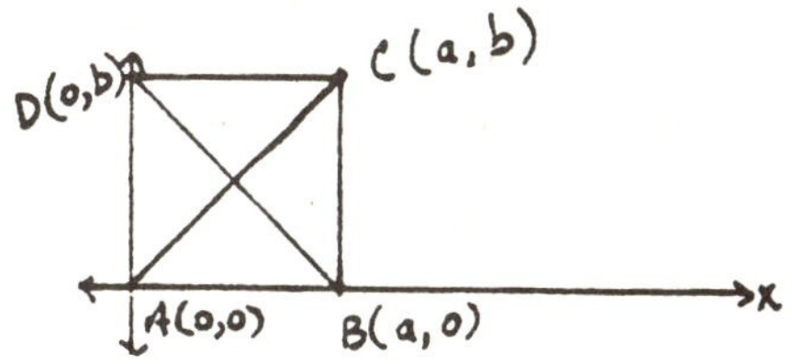

Proof: Let rect. ABCD have vertices as shown.

If $\overline{DB} \perp \overline{AC}$, then $m_{\overline{DB}} \cdot m_{\overline{AC}} = -1$.

But $m_{\overline{DB}} = \frac{0 - b}{a - 0} = \frac{-b}{a}$ and

$m_{\overline{AC}} = \frac{b - 0}{a - 0} = \frac{b}{a}$

Then $\frac{-b}{a} \cdot \frac{b}{a} = -1 \rightarrow \frac{-b^2}{a^2} = -1$

$\rightarrow -b^2 = -a^2 \rightarrow b^2 = a^2$.

Since a and b are both positive, a = b.

Then AB = a − 0 = a and

AD = b − 0 = b. Since a = b, AB = AD.

If AB = AD, then ABCD is a square.

34. If the diagonals of a trapezoid are equal in length, then the trapezoid is an isosceles trapezoid.

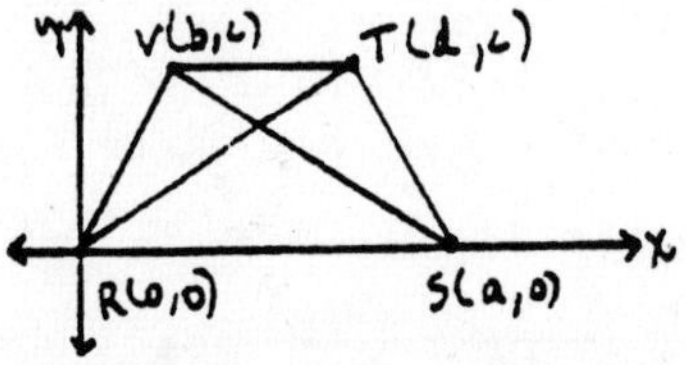

Proof: Let trap. RSTV have vertices as shown.
If RT = VS, then

$$\sqrt{(d-0)^2+(c-0)^2}$$

$$=\sqrt{(a-b)^2+(0-c)^2}$$

$$\sqrt{d^2+c^2}=\sqrt{(a-b)^2+(-c)^2}$$

$$\sqrt{d^2+c^2}=\sqrt{(a-b)^2+c^2}$$

Squaring, $d^2+c^2=(a-b)^2+c^2$

$\therefore d^2=(a-b)^2$

$d=a-b$ (*)

Comparing lengths of $\overline{RV}$ and $\overline{ST}$,

$$RV=\sqrt{(b-0)^2+(c-0)^2}=\sqrt{b^2+c^2}$$

and

$$ST=\sqrt{(a-d)^2+(0-c)^2}=\sqrt{(a-d)^2+c^2}$$

Now RV would equal ST if

$b^2+c^2=(a-d)^2+c^2$

$b^2=(a-d)^2$

or $b=a-d$

Because $d=a-b$ leads to $b=a-d$, so RV = ST. Then RSTV is isosceles.

35. If two medians of a triangle are equal in length, then the triangle is isosceles.

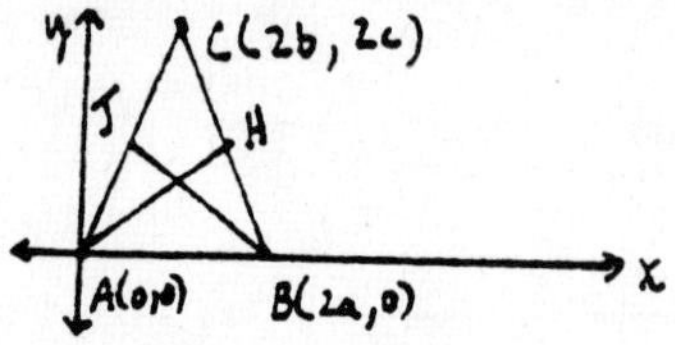

Proof: Let $\triangle ABC$ have vertices as shown, so that the midpoint of $\overline{AC}$ is J = (b,c) and the midpoint of $\overline{BC}$ is H = (a + b,c).
If AH = BJ, then

$$\sqrt{[(a+b)-0]^2+(c-0)^2}$$

$$=\sqrt{(2a-b)^2+(0-c)^2}$$

$$\sqrt{(a+b)^2+c^2}=\sqrt{(2a-b)^2+c^2}$$

Squaring, $(a+b)^2+c^2=(2a-b)^2+c^2$

$(a+b)^2=(2a-b)^2$

Taking the principal square roots,

$a+b=2a-b$

$2b=a$

Then point C = (a,2c).

Now $AC=\sqrt{(a-0)^2+(2c-0)^2}$

$=\sqrt{a^2+(2a)^2}=\sqrt{a^2+4c^2}$

and $BC=\sqrt{(2a-a)^2+(0-2c)^2}$

$=\sqrt{a^2+(-2c)^2}$

$=\sqrt{a^2+4c^2}$

Because AC = BC, the triangle is isosceles.

36. The segments joining the midpoints of consecutive sides of an isosceles trapezoid form a rhombus.

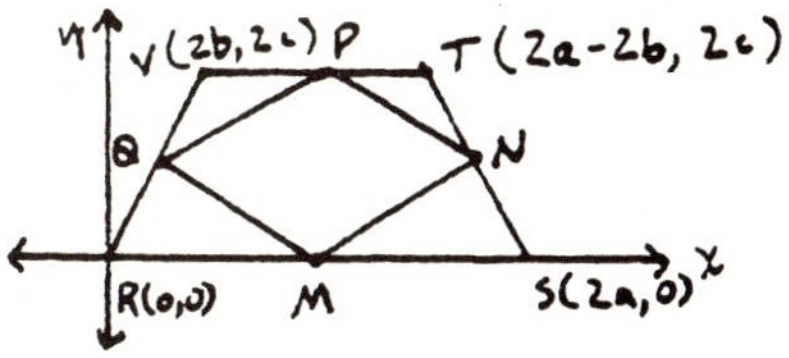

Proof: Let trapezoid RSTV have vertices as shown so that RV = TS.

Where M, N, P, and Q are the midpoints of the sides,

$$M = \left(\frac{0 + 2a}{2}, \frac{0 + 0}{2}\right) = (a,0)$$

$$N = \left(\frac{(2a - 2b) + 2a}{2}, \frac{0 + 2c}{2}\right) = (2a - b,c)$$

$$P = \left(\frac{(2a - 2b) + 2b}{2}, \frac{2c + 2c}{2}\right) = (a,2c)$$

$$Q = \left(\frac{0 + 2b}{2}, \frac{0 + 2c}{2}\right) = (b,c)$$

By an earlier theorem, MNPQ is a parallelogram.

We must show that two adjacent sides of parallelogram MNPQ are equal in length.

$$QM = \sqrt{(a - b)^2 + (0 - c)^2} = \sqrt{(a - b)^2 + (-c)^2} = \sqrt{(a - b)^2 + c^2}$$

$$MN = \sqrt{[(2a - b) - a]^2 + (c - 0)^2} = \sqrt{(a - b)^2 + c^2}.$$

Now QM = MN, so MNPQ is a rhombus.

Chapter Ten: Introduction to Trigonometry

SECTION 10.1: The Sine Ratio and Applications

1. $\sin \alpha = \frac{\text{opposite}}{\text{hypotenuse}} = \frac{5}{13}$;

   $\sin \beta = \frac{12}{13}$

5. $a^2 + b^2 = c^2$

   $(\sqrt{2})^2 + (\sqrt{3})^2 = c^2$

   $2 + 3 = c^2$

   $c^2 = 5$

   $c = \sqrt{5}$

   $\sin \alpha = \frac{\sqrt{3}}{\sqrt{5}} = \frac{\sqrt{3}}{\sqrt{5}} \cdot \frac{\sqrt{5}}{\sqrt{5}} = \frac{\sqrt{15}}{5}$;

   $\sin \beta = \frac{\sqrt{2}}{\sqrt{5}} = \frac{\sqrt{2}}{\sqrt{5}} \cdot \frac{\sqrt{5}}{\sqrt{5}} = \frac{\sqrt{10}}{5}$

9. $\sin 17^\circ = 0.2924$

13. $\sin 72^\circ = 0.9511$

17. $\sin 43^\circ = \frac{a}{16}$

    $a = 16 \sin 43^\circ$

    $a \approx 16(0.6820)$

    $a \approx 10.9$

    $\sin 47^\circ = \frac{b}{16}$

    $b = 16 \sin 47^\circ$

    $b \approx 16(0.7314)$

    $b \approx 11.7'$

21. $\sin \alpha = \frac{12}{25} = 0.4800$

    $\alpha \approx 29^\circ$

    $\beta \approx 90^\circ - 29^\circ$ or $\beta \approx 61^\circ$

25. $\sin \alpha = \frac{x}{3x} = \frac{1}{3} \approx 0.3333$

    $\alpha \approx 19^\circ$

    $\beta \approx 90^\circ - 19^\circ$ or $\beta \approx 71^\circ$

29. Let d represent the distance between Danny and the balloon.

    $\sin 75^\circ = \frac{100}{d}$

    $d \cdot \sin 75^\circ = 100$

    $d = \frac{100}{\sin 75^\circ}$

    $d \approx \frac{100}{0.9659}$

    $d \approx 103.5$ ft.

33. $\sin \alpha = \frac{4}{10}$

    $\sin \alpha = 0.4000$

    $\alpha \approx 24^\circ$

37.

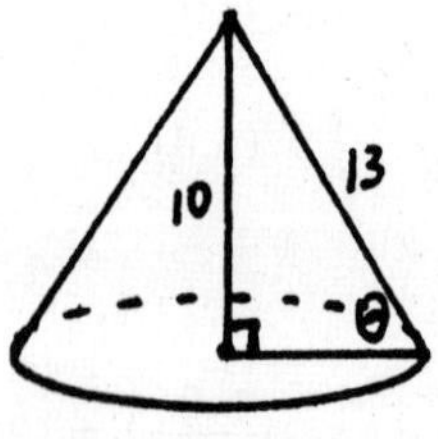

$\sin \theta = \frac{10}{13} \approx 0.7692$

$\theta \approx 50^\circ$

SECTION 10.2: The Cosine Ratio and Applications

1. $\cos \alpha = \frac{\text{adjacent}}{\text{hypotenuse}} = \frac{12}{13}$; $\cos \beta = \frac{5}{13}$

5. $a^2 + b^2 = c^2$

   $(\sqrt{3})^2 + (\sqrt{2})^2 = c^2$

   $c^2 = 3 + 2$

   $c^2 = 5$

   $c = \sqrt{5}$

   $\cos \alpha = \frac{\sqrt{2}}{\sqrt{5}} = \frac{\sqrt{2}}{\sqrt{5}} \cdot \frac{\sqrt{5}}{\sqrt{5}} = \frac{\sqrt{10}}{5}$,

   $\cos \beta = \frac{\sqrt{3}}{\sqrt{5}} = \frac{\sqrt{3}}{\sqrt{5}} \cdot \frac{\sqrt{5}}{\sqrt{5}} = \frac{\sqrt{15}}{5}$.

9. $\cos 23^\circ \approx 0.9205$

13. $\cos 90^\circ = 0$

17. $\cos 32^\circ = \frac{a}{100}$

$a = 100 \cos 32^\circ$

$a \approx 100(0.8480)$

$a \approx 84.8'$

$\sin 32^\circ = \frac{b}{100}$

$b = 100 \sin 32^\circ$

$b \approx 100(0.5299)$

$b \approx 53.0'$

21. $\cos 51^\circ = \frac{12}{c}$

$c \cdot \cos 51^\circ = 12$

$c = \frac{12}{\cos 51^\circ}$

$c \approx \frac{12}{0.6293}$

$c \approx 19.1''$

$\sin 51^\circ = \frac{d}{c}$

$d = c \cdot 51^\circ$

$d \approx (19.1)(0.7771)$

$d \approx 14.8''$

25. $a^2 + b^2 = c^2$

$(\sqrt{3})^2 + (\sqrt{2})^2 = c^2$

$c^2 = 3 + 2$

$c^2 = 5$

$c = \sqrt{5}$

$\cos \alpha = \frac{\sqrt{2}}{\sqrt{5}} = \sqrt{\frac{2}{5}} = \sqrt{0.4} \approx 0.6325$

$\alpha \approx 51^\circ$

$\beta \approx = 90^\circ - 51^\circ$

$\beta \approx 39^\circ$

29. $\cos \theta = \frac{10}{12} \approx 0.8333$

$\theta \approx 34^\circ$

33. Let c represent the measure of the central angle of the regular pentagon. Then $c = \frac{360^\circ}{5} = 72^\circ$.

Because the apothem shown bisects the central angle,

$\cos 36^\circ = \frac{a}{10}$

$a = 10 \cos 36^\circ$

$a \approx 10\ (0.8090)$

$a \approx 8.1''$

37. Let d represent the length of the diagonal of the base and x the length of an edge of the cube. By the Pythagorean Theorem,

$d^2 = x^2 + x^2$

$d^2 = 2x^2.$

Let D represent the length of the diagonal of the cube.

Applying the Pythagorean Theorem again,

$x^2 + d^2 = D^2$

$x^2 + 2x^2 = D^2$

$D^2 = 3x^2$

$D = \sqrt{3x^2}$

$D = x\sqrt{3}$

Now $\cos \alpha = \frac{\text{adjacent}}{\text{hypotenuse}} = \frac{x}{x\sqrt{3}} = \frac{1}{\sqrt{3}} \approx 0.5774$

$\alpha \approx 55^\circ$

41.

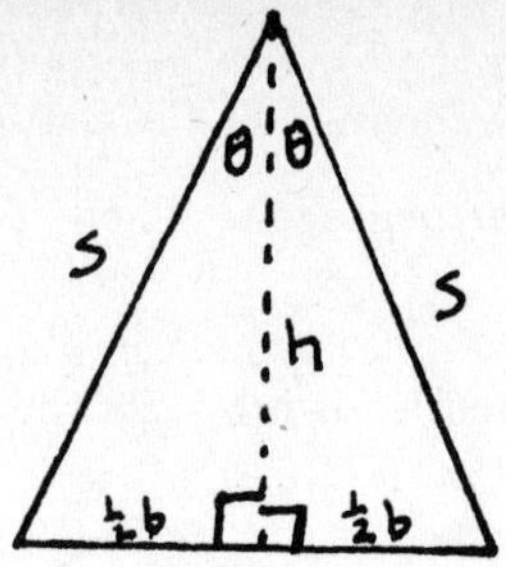

$\sin\theta = \frac{\frac{1}{2}b}{s}$

$s \cdot \sin\theta = \frac{1}{2}b$

$b = 2s \cdot \sin\theta$

$\cos\theta = \frac{h}{s}$

$h = s \cdot \cos\theta$

$A_{\triangle} = \frac{1}{2}bh$

$A_{\triangle} = \frac{1}{2}(2 \cdot \sin\theta)(s \cdot \cos\theta)$

$A_{\triangle} = s^2 \cdot \sin\theta \cdot \cos\theta$

SECTION 10.3: The Tangent Ratio and Other Ratios

1. $\tan\alpha = \frac{\text{opposite}}{\text{adjacent}} = \frac{3}{4};$

$\tan\beta = \frac{4}{3}$

5. Using the Pythagorean Triple, (5, 12, 13), b = 12.

$\sin\alpha = \frac{\text{opposite}}{\text{hypotenuse}} = \frac{5}{13}$

$\cos\alpha = \frac{\text{adjacent}}{\text{hypotenuse}} = \frac{12}{13}$

$\tan\alpha = \frac{\text{opposite}}{\text{adjacent}} = \frac{5}{12}$

$\cot\alpha = \frac{\text{adjacent}}{\text{opposite}} = \frac{12}{5}$

$\sec\alpha = \frac{\text{hypotenuse}}{\text{adjacent}} = \frac{13}{12}$

$\csc\alpha = \frac{\text{hypotenuse}}{\text{opposite}} = \frac{13}{5}$

9. $a^2 + b^2 = c^2$

$x^2 + b^2 = (\sqrt{x^2+1})^2$

$x^2 + b^2 = x^2 + 1$

$b^2 = 1$

$b = 1$

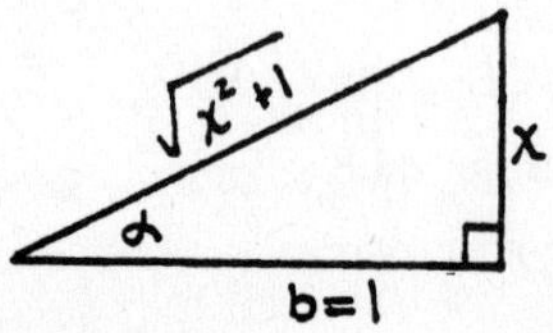

$\sin\alpha = \frac{x}{\sqrt{x^2+1}}$

$= \frac{x}{\sqrt{x^2+1}} \cdot \frac{\sqrt{x^2+1}}{\sqrt{x^2+1}}$

$= \frac{x\sqrt{x^2+1}}{x^2+1}$

$\cos\alpha = \frac{1}{\sqrt{x^2+1}}$

$= \frac{1}{\sqrt{x^2+1}} \cdot \frac{\sqrt{x^2+1}}{\sqrt{x^2+1}}$

$= \frac{\sqrt{x^2+1}}{x^2+1}$

$\tan\alpha = \frac{x}{1}$

$\cot\alpha = \frac{1}{x}$

$\sec\alpha = \frac{\sqrt{x^2+1}}{1} = \sqrt{x^2+1}$

$\csc\alpha = \frac{\sqrt{x^2+1}}{x}$

13. $\tan 57^\circ = 1.5399$

17. $\cos 58^\circ = \frac{y}{10}$

$y = 10 \cos 58^\circ$

$y \approx 10(0.5299)$

$y \approx 5.3$

$\sin 58^\circ = \frac{z}{10}$

$z = 10 \sin 58^\circ$

$z \approx 10(0.8480)$

$z \approx 8.5$

21. $\tan \alpha = \frac{3}{4} = 0.7500$

$\alpha \approx 37^\circ$

$\beta \approx 90^\circ - 37^\circ$

$\beta \approx 53^\circ$

25. $\tan \alpha = \frac{\sqrt{5}}{4} \approx \frac{2.2361}{4} \approx 0.5590$

$\alpha \approx 29^\circ$

$\beta \approx 90^\circ - 29^\circ$

$\beta \approx 61^\circ$

29. $\csc 30^\circ = \frac{1}{\sin 30^\circ} = \frac{1}{0.5} = 2.0000$ (exact)

33. $\sin 5^\circ = \frac{120}{x}$

$x \cdot \sin 5^\circ = 120$

$x = \frac{120}{\sin 5^\circ}$

$x \approx \frac{120}{0.0872} \approx 1{,}376.8$ ft.

37. $\cos \alpha = \frac{10}{32}$

$\cos \alpha = 0.3125$

$\alpha \approx 72^\circ$

41.

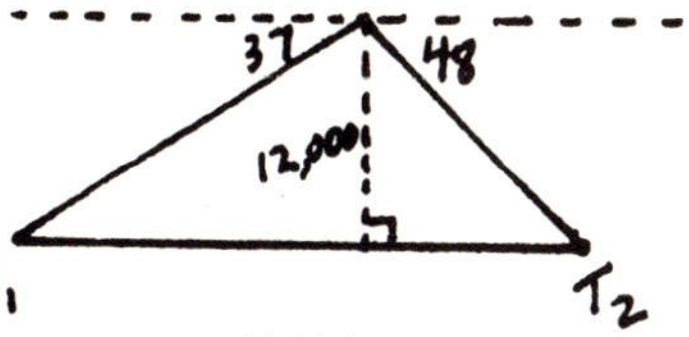

$\tan 37 = \frac{12000}{x}$

$x \cdot \tan 37 = 12000$

$x = \frac{12000}{\tan 37}$

$x \approx 15{,}924.5$

$\tan 48^\circ = \frac{12000}{y}$

$y \cdot \tan 48 = 12000$

$y = \frac{12000}{\tan 48}$

$y \approx 10{,}804.8$

Distance $\approx 15{,}924.5 + 10{,}804.8$

$\approx 26{,}729.3 \approx 26{,}730$ feet

SECTION 10.4: More Trigonometric Relationships

1 . $\frac{\cos \theta}{\sin \theta} = \frac{\frac{b}{c}}{\frac{a}{c}} = \frac{b}{c} \cdot \frac{c}{a} = \frac{b}{a} = \cot \theta$

5 . $\sin^2\theta + \cos^2\theta = 1$

$\left(\frac{3}{5}\right)^2 + \cos^2\theta = 1$

$\frac{9}{25} + \cos^2\theta = 1$

$\cos^2\theta = \frac{16}{25}$

$\cos \theta = \sqrt{\frac{16}{25}} = \frac{\sqrt{16}}{\sqrt{25}} = \frac{4}{5}.$

9. Show that $\cot^2\theta + 1 = \csc^2\theta$

Using the rt. Δ, $\cot\theta = \frac{b}{a}$ while

$\csc\theta = \frac{c}{a}$

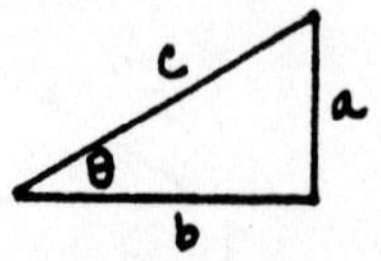

Now $\cot^2\theta + 1 = \left(\frac{b}{a}\right)^2 + 1$

$$= \frac{b^2}{a^2} + 1$$

$$= \frac{b^2}{a^2} + \frac{a^2}{a^2}$$

$$= \frac{a^2 + b^2}{a^2}$$

$$= \frac{c^2}{a^2}$$

$$= \left(\frac{c}{a}\right)^2$$

$$= \csc^2\theta$$

13. With measures of angles shown, the third angle measures 70°. Then the triangle is isosceles, with sides as shown.

$A = \frac{1}{2}\cdot 6\cdot 6\cdot \sin 40°$

$A \approx \frac{1}{2}\cdot 6\cdot 6\cdot (0.6428)$

$A \approx 11.6 \text{ ft}^2$

17. $\frac{\sin\alpha}{a} = \frac{\sin\beta}{b}$

$\frac{\sin 60°}{x} = \frac{\sin 70°}{12}$

$\frac{0.8660}{x} = \frac{0.9397}{12}$

$x(0.9397) = 12(0.8660)$

$x \approx 11.1$ in.

21. $\frac{\sin\gamma}{10} = \frac{\sin 80°}{12}$

$\frac{\sin\gamma}{10} = \frac{0.9848}{12}$

$12(\sin\gamma) = 10(0.9848)$

$\sin\gamma \approx 0.8207$

$\gamma \approx 55°$

25. $x^2 = 8^2 + 12^2 - 2\cdot 8\cdot 12\cos 60°$

$x^2 = 64 + 144 - 192\cos 60°$

$x^2 = 208 - 192(0.5)$

$x^2 = 208 - 96$

$x^2 = 112$

$x = \sqrt{112} \approx 10.6$

29. a. $x^2 = 150^2 + 180^2 - 2(150)(180)\cos 80°$

$x^2 = 22500 + 32400 - 54000\ (0.1736)$

$x^2 = 44525.6$

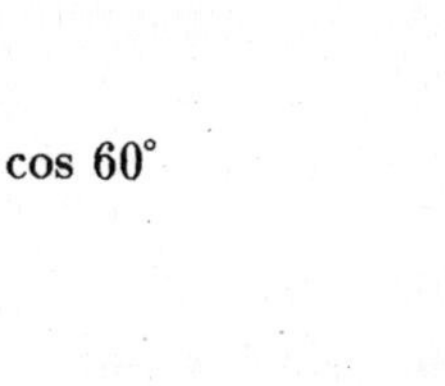

$x = \sqrt{44525.6}$

$x \approx 213.4$ feet

b. $A = \frac{1}{2}(150)(180)\sin 80^\circ$

$A = \frac{1}{2}(150)(180)(0.9848)$

$A \approx 13{,}294.9 \text{ ft}^2$

33. The third angle measures 85°.

Using the Law of Sines,

$\frac{\sin 85^\circ}{x} = \frac{\sin 30^\circ}{8}$

$\frac{0.9962}{x} = \frac{0.5}{8}$

$0.5x = 8(0.9962)$

$x \approx 15.9$ ft.

37. $a^2 = 27^2 + 27^2 - 2(27)(27)\cos 30$

$a^2 = 729 + 729 - 2(729)\cdot\frac{\sqrt{3}}{2}$

$a^2 \approx 195.33$

$a \approx 13.97 \approx 14.0$ feet

41. $A = (6.3)(8.9)\sin 67.5 = 51.8 \text{ cm}^2$

CHAPTER 10: Review

1. sine;

$\sin 40^\circ = \frac{a}{16}$

$a = 16 \sin 40$

$a \approx 10.3$ in

2. sine;

$\sin 70^\circ = \frac{d}{8}$

$d = 8 \sin 70$

$d \approx 7.5$ ft

3. cosine;

$\cos 80^\circ = \frac{4}{c}$

$c = \frac{4}{\cos 80} \approx \frac{4}{0.1736}$

$c \approx 23$ in

4 . sine:

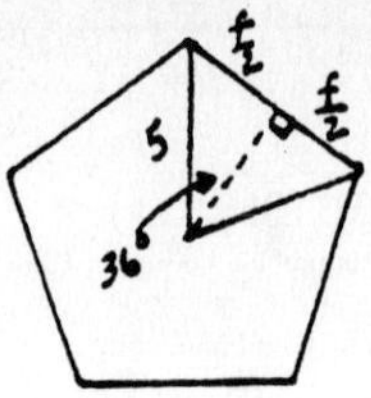

$\sin 36^\circ = \dfrac{\frac{1}{2}f}{5}$

$\frac{1}{2}f = 5(\sin 36^\circ)$

$\frac{1}{2}f \approx 5(0.5878)$

$\frac{1}{2}f \approx 2.939$

$f \approx 5.9$ ft.

5 . tangent;

$\tan \alpha = \dfrac{13}{14}$

$\tan \alpha \approx 0.9286$

$\alpha \approx 43^\circ$

6 . cosine;

$\cos \theta = \dfrac{8}{15}$

$\cos \theta \approx 0.5333$

$\theta \approx 58^\circ$

7 . sine;

$\sin \alpha = \dfrac{9}{12}$

$\sin \alpha = 0.7500$

$\alpha \approx 49^\circ$

8 . tangent;

$\tan \beta = \dfrac{7}{24}$

$\tan \beta \approx 0.2917$

$\beta \approx 16^\circ$

9 . Law of Sines

$\dfrac{\sin 57^\circ}{x} = \dfrac{\sin 49^\circ}{8}$

$x(\sin 49^\circ) = 8(\sin 57^\circ)$

$x(0.7547) = 8(0.8387)$

$x \approx 8.9$ units

10. Law of Cosines

$15^2 = 14^2 + 16^2 - 2(14)(16) \cos \alpha$

$225 = 196 + 256 - 448 \cos \alpha$

$448 \cos \alpha = 227$

$\cos \alpha = \dfrac{227}{448} \approx 0.5067$

$\alpha \approx 60^\circ$

11. The third angle measures 80°.

Law of Sines

$\dfrac{\sin 40^\circ}{y} = \dfrac{\sin 80^\circ}{20}$

$y(\sin 80^\circ) = 20(\sin 40^\circ)$

$y(0.9848) = 20(0.6428)$

$y \approx 13.1$

12. Law of Cosines

$w^2 = 14^2 + 21^2 - 2(14)(21) \cos 60^\circ$

$w^2 = 196 + 441 - 294$

$w^2 = 343$

$w = \sqrt{343} \approx 18.5$

13. The remaining angles of the triangle measure 47° and 74°.

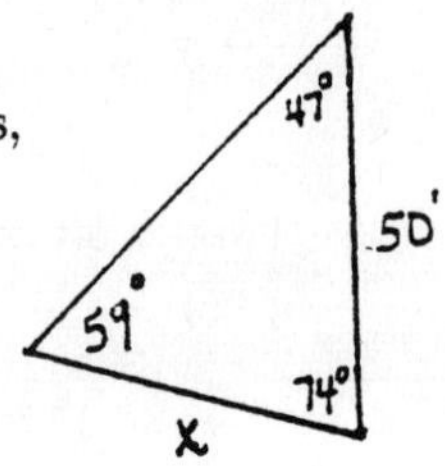

Using the Law of Sines,

$\dfrac{\sin 47^\circ}{x} = \dfrac{\sin 59^\circ}{50}$

$\dfrac{0.7314}{x} = \dfrac{0.8572}{50}$

$x(0.8572) = 50(0.7314)$

$x \approx 42.7$ feet

14. Let d be the length of the shorter diagonal.

Law of Cosines

$d^2 = 50^2 + 70^2 - 2(50)(70) \cos 75^\circ$

$d^2 = 2500 + 4900 - 7000(0.2588)$

$d^2 = 2500 + 4900 - 1811.6$

$d^2 = 5588.4$

$d = \sqrt{5588.4} \approx 74.8$ cm

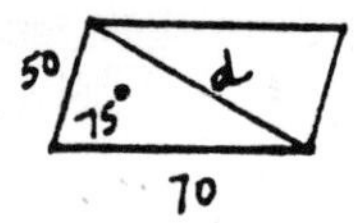

15. Law of Cosines

$6^2 = 6^2 + 11^2 - 2(6)(11) \cos \alpha$

$36 = 36 + 121 - 132 \cos \alpha$

$132 \cos \alpha = 121$

$\cos \alpha = \dfrac{121}{132} \approx 0.9167$

$\alpha = 23.6^\circ$

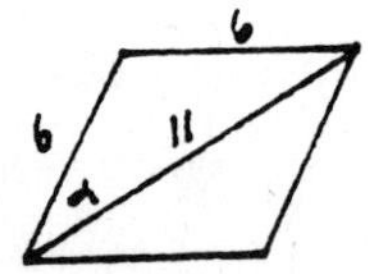

The acute angle of the rhombus measures

$2\alpha \approx 47^\circ$

16. $A = \frac{1}{2}\,ac \sin B$

$9.7 = \frac{1}{2}(6)(4) \sin B$

$9.7 = 12 \sin B$

$\sin B = \frac{9.7}{12} \approx 0.8083$

$B \approx 54^\circ$

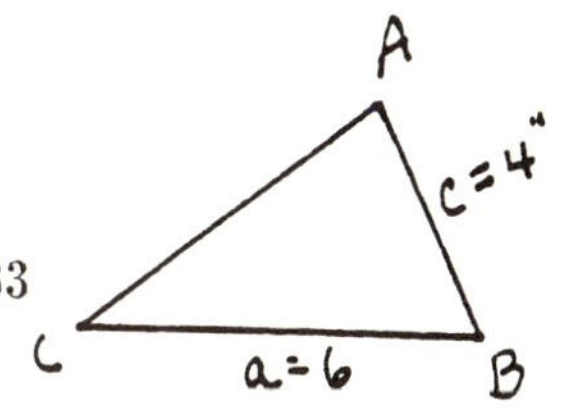

17. If the acute angle measures 47°, then

$A = \frac{1}{2}\,ab \sin 47^\circ$ gives one-half the desired area.

$A = \frac{1}{2} \cdot 6 \cdot 6 \sin 47^\circ$

$A \approx 13.16 \text{ in}^2$ for $\triangle ABC$.

The area of rhombus ABCD is approximately 26.3 in^2.

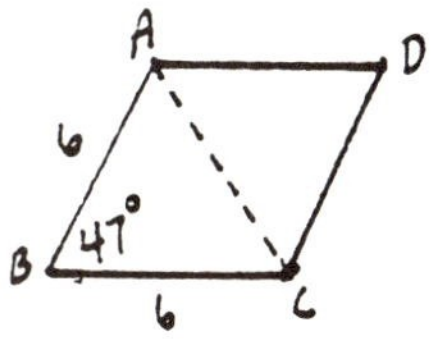

18. If $m\angle R = 45^\circ$, then $m\angle S = 45^\circ$ also.

Then $\overline{RT} \doteq \overline{ST}$.

Let $RT = ST = x$.

Now $\tan R = \tan 45^\circ$

$= \frac{x}{x}$

$= 1.$

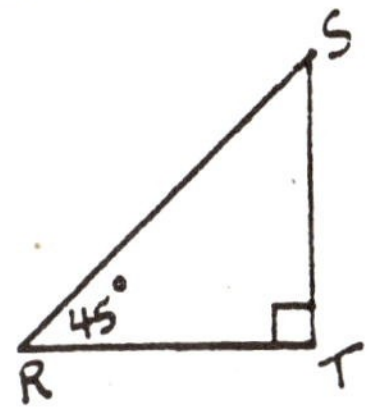

19. If $m\angle S = 30$, then the sides of $\triangle RQS$ can be represented by $RQ = x$, $RS = 2x$, and $SQ = x \cdot \sqrt{3}$.

$\sin S = \sin 30^\circ = \frac{x}{2x} = \frac{1}{2}.$

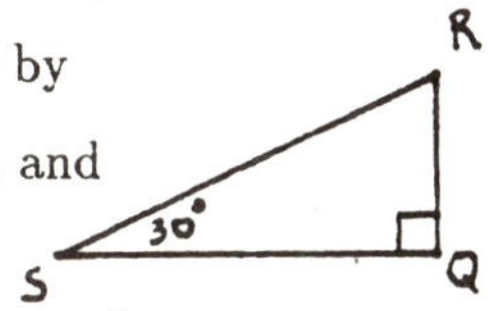

20. If $m\angle T = 60^\circ$, then the sides of $\triangle TVW$ can be represented by

$TV = x$,

$TW = 2x$, and

$VW = x\sqrt{3}$.

$\sin T = \sin 60^\circ = \frac{x\sqrt{3}}{2x} = \frac{\sqrt{3}}{2}.$

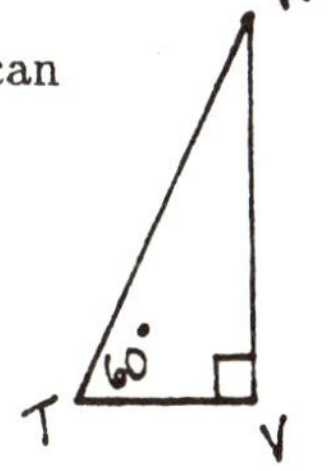

21. Because alt. int. $\angle$s are $\doteq$,

$\tan 55^\circ = \frac{12}{x}$

$x(\tan 55^\circ) = 12$

$x = \frac{12}{\tan 55^\circ} \approx \frac{12}{1.4281}$

$x \approx 8.4$ ft.

22. $\sin 60^\circ = \frac{x}{200}$

$x = 200 \sin 60^\circ$

$x \approx 200\,(0.866)$

$x \approx 173.2$

If the rocket rises 173.2 feet per second, then its altitude after 5 seconds will be approximately 866 feet.

23. $\cos \alpha = \frac{3}{4}$

$\cos \alpha = 0.75$

$\alpha \approx 41^\circ$

24. $\sin \alpha = \frac{300}{2200}$

$\sin \alpha = 0.1364$

$\alpha \approx 8^\circ$

25. Let x represent one-half the length of a side of the regular pentagon.

$\tan 36^\circ = \frac{x}{3.44}$

$x = 3.44\,(\tan 36^\circ)$

$x \approx 3.44(0.7265)$

$x \approx 2.50$

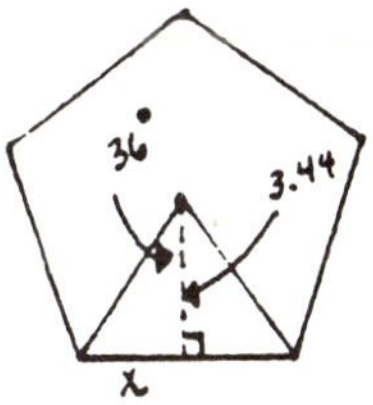

Then the length of each side is approximately 5.0 cm.

26. $\cos 36^\circ = \frac{3.44}{r}$

$r \cos 36^\circ = 3.44$

$r = \frac{3.44}{\cos 36^\circ}$

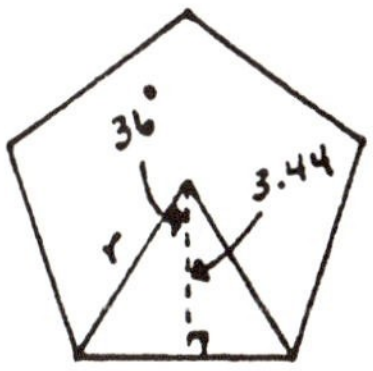

$r \approx \frac{3.44}{0.8090} \approx 4.3$ cm

27. The altitude bisects the base where β represents the measure of the base angle.

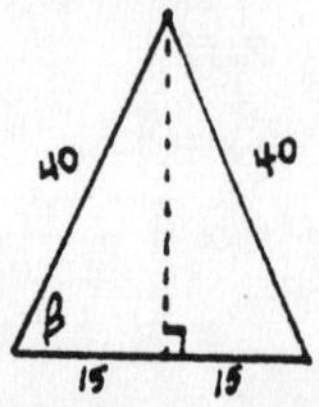

$\cos \beta = \frac{15}{40}$

$\cos \beta = 0.375$

$\beta \approx 68^\circ$

28. The measure of acute angle α is one-half the desired angle's measure.

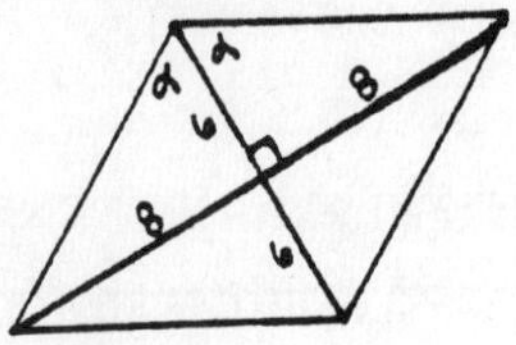

$\tan \alpha = \frac{8}{6}$

$\tan \alpha \approx 1.333$

$\alpha \approx 53^\circ$

The obtuse angle of the rhombus measures approximately 106°.

29. $\tan 23^\circ = \frac{a}{b}$

$\tan 23^\circ = \frac{3}{b}$

$b(\tan 23^\circ) = 3$

$b = \frac{3}{\tan 23^\circ}$

$b \approx \frac{3}{0.4245} \approx 7$

The grade of the hill is 3 to 7 (or 3:7).

30. Let S_1 represent the distance to the nearer ship and S_2 the distance to the farther ship.

$\tan 32^\circ = \frac{2500}{S_2}$ and $\tan 44^\circ = \frac{2500}{S_1}$

$S_2(\tan 32^\circ) = 2500 \qquad S_1(\tan 44^\circ) = 2500$

$S_2 = \frac{2500}{\tan 32^\circ} \qquad S_1 = \frac{2500}{\tan 44^\circ}$

$S_2 \approx 4000.8 \qquad S_1 \approx 2588.8$

The distance between the ships is $4000.8 - 2588.8$ or approximately 1,412.0 meters.

31. $\sin \theta = \frac{7}{25}$

Using $\sin^2\theta + \cos^2\theta = 1$,

$\left(\frac{7}{25}\right)^2 + \cos^2\theta = 1 \rightarrow \frac{49}{625} + \cos^2\theta = 1$

$\rightarrow \cos^2\theta = 1 - \frac{49}{625} \rightarrow \cos^2\theta = \frac{576}{625}$

$\rightarrow \cos \theta = \sqrt{\frac{576}{625}} = \frac{\sqrt{576}}{\sqrt{625}} = \frac{24}{25}$

Because $\sec \theta = \frac{1}{\cos \theta}$, $\sec \theta = \frac{1}{\frac{24}{25}} = \frac{25}{24}$.

32. $\tan \theta = \frac{11}{60}$

Using $\tan^2\theta + 1 = \sec^2\theta$

$\left(\frac{11}{60}\right)^2 + 1 = \sec^2\theta \rightarrow \frac{121}{3600} + 1 = \sec^2\theta \rightarrow$

$\frac{3721}{3600} = \sec^2\theta$

$\sec \theta = \sqrt{\frac{3721}{3600}} = \frac{\sqrt{3721}}{\sqrt{3600}} = \frac{61}{60}$.

Because $\cot \theta = \frac{1}{\tan \theta}$, $\cot \theta = \frac{1}{\frac{11}{60}} = \frac{60}{11}$.

33. $\cot \theta = \frac{21}{20}$

Using $\cot^2\theta + 1 = \csc^2\theta$

$\left(\frac{21}{20}\right)^2 + 1 = \csc^2\theta$

$\frac{441}{400} + 1 = \csc^2\theta$

$\frac{841}{400} = \csc^2\theta$

$\csc \theta = \sqrt{\frac{841}{400}} = \frac{\sqrt{841}}{\sqrt{400}} = \frac{29}{20}$

Because $\sin \theta = \frac{1}{\csc \theta}$,

$\sin \theta = \frac{1}{\frac{29}{20}} = \frac{20}{29}$.

34.

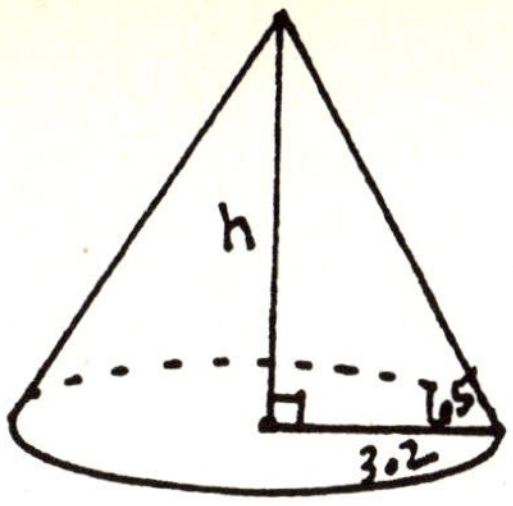

$\tan 65^\circ = \frac{h}{3.2}$

$h = 3.2 \cdot \tan 65$

$h \approx 6.9$ feet

$V = \frac{1}{3}Bh$

$V = \frac{1}{3}\pi r^2 h$

$V \approx \frac{1}{3}\pi \cdot (3.2)^2 \cdot (6.9)$

$V \approx 74.0 \text{ ft}^3$

Appendix A

A.1: Algebraic Expressions

1. Undefined terms, definitions, axioms or postulates, theorems.
2. Algebra; geometry
3. a. Reflexive
   b. Transitive
   c. Substitution
   d. Symmetric
4. a. $2 = 2$
   b. If $2 = x$, then $x = 2$.
   c. If $x = 2$ and $2 = y$, then $x = y$.
   d. If $2 + 5 = 7$ and $7 + y = z$, then $2 + 5 + y = z$.
5. a. 12
   b. -2
   c. 2
   d. -12
6. a. 8
   b. -8
   c. -22
   d. 1
7. a. 35
   b. -35
   c. -35
   d. 35
8. a. -84
   b. 84
   c. -84
   d. 84
9. No; Commutative Axiom for Multiplication
10. a. Commutative Axiom for Multiplication
    b. Associative Axiom for Addition
    c. Commutative Axiom for Addition
    d. Associative Axiom for Multiplication
11. a. 9
    b. -9
    c. 8
    d. -8
12. $(-3) - 7$
13. a. -4
    b. -36
    c. 18
    d. $-\frac{1}{4}$
14. 5 feet divided by 10 spaces $= \frac{1}{2}$ ft. per space. Or since 5 feet $= 60$ inches, 60 inches divided by 10 spaces $= 6$ inches per space.
15. $-\$60$
16. $25(2) + 30(2) = 50 + 60 = \$110$
17. a. $30 + 35 = 65$
    b. $28 - 12 = 16$
    c. $\frac{7}{2} + \frac{11}{2} = \frac{18}{2} = 9$
    d. $8x$
18. a. $54 - 24 = 30$
    b. $3(4 + 8) = 12 + 24 = 36$
    c. $7y - 2y = (7 - 2)y = 5y$
    d. $(16 + 8)x = 24x$
19. a. $(6 + 4)\pi = 10\pi$
    b. $(8 + 3)\sqrt{2} = 11\sqrt{2}$
    c. $(16 - 9)x^2y = 7x^2y$
    d. $(9 - 2)\sqrt{3} = 7\sqrt{3}$
20. a. $(1 + 2)\pi r^2 = 3\pi r^2$
    b. $(7 + 3)xy = 10xy$
    c. $7x^2y + 3xy^2$
    d. $(1 + 1)x + y = 2x + y$

21. a. $2 + 12 = 14$

b. $5 \cdot 4 = 20$

c. $2 + 3 \cdot 4 = 2 + 12 = 14$

d. $2 + 6^2 = 2 + 36 = 38$

22. a. $9 + 16 = 25$

b. $7^2 = 49$

c. $9 + 6 \div 3 = 9 + 2 = 11$

d. $[9 + 6] \div 3 = 15 \div 3 = 5$

23. a. $\frac{6}{-6} = -1$

b. $\frac{8-6}{6 \cdot 3} = \frac{2}{18} = \frac{1}{9}$

c. $\frac{10-18}{9} = \frac{-8}{9}$

d. $\frac{5 - 12 + (-3)}{4 + 16} = \frac{-10}{20} = \frac{-1}{2}$

24. a. $8 + 10 + 12 + 15 = 45$

b. $42 + 7 - 12 - 2 = 35$

25. a. $15 - 6 - 5 + 2 = 6$

b. $12x^2 - 15x + 8x - 10$

$= 12x^2 - 7x - 10$

26. a. $10x^2 - 35x + 6x - 21$

$= 10x^2 - 29x - 21$

b. $6x^2 - 10xy + 3xy - 5y^2$

$= 6x^2 - 7xy - 5y^2$

27. $5x + 2y$

28. $2xy + 2yz + 2xz$

29. $10x + 5y$

30. $xy + xz + y^2 + yz$; the total of the areas of the four smaller plots is also $xy + y^2 + xz + yz$.

31. $10x$

32. $9\pi + 48\pi + 9\pi = 66\pi$

A.2: Formulas and Equations

1. $5x + 8$

2. $-1x + 8$

3. $2x - 2$

4. $4x - 2$

5. $2x + 2 + 3x + 6 = 5x + 8$

6. $6x + 15 - 6x + 2 = 17$

7. $x^2 + 4x + 3x + 12 = x^2 + 7x + 12$

8. $x^2 - 7x - 5x + 35 = x^2 - 12x + 35$

9. $6x^2 - 4x + 15x - 10 = 6x^2 + 11x - 10$

10. $6x^2 + 9x + 14x + 21 = 6x^2 + 23x + 21$

11. $(a + b)(a + b) + (a - b)(a - b)$

$a^2 + ab + ab + b^2 + a^2 - ab - ab + b^2$

$2a^2 + 2b^2$

12. $(x + 2)(x + 2) - (x - 2)(x - 2)$

$x^2 + 2x + 2x + 4 - (x^2 - 2x - 2x + 4)$

$x^2 + 4x + 4 - (x^2 - 4x + 4)$

$x^2 + 4x + 4 - x^2 + 4x - 4$

$8x$

13. $4 \cdot 3 \cdot 5 = 60$

14. $5^2 + 7^2 = 25 + 49 = 74$

15. $2 \cdot 13 + 2 \cdot 7 = 26 + 14 = 40$

16. $6 \cdot 16 \div 4 = 96 \div 4 = 24$

17. $S = 2 \cdot 6 \cdot 4 + 2 \cdot 4 \cdot 5 + 2 \cdot 6 \cdot 5$

$S = 48 + 40 + 60$

$S = 148$

18. $A = \left(\frac{1}{2}\right)2(6 + 8 + 10)$

$A = 1(24)$

$A = 24$

19. $V = \left(\frac{1}{3}\right)\pi(3)^2 \cdot 4$

$V = \frac{1}{3} \cdot \pi \cdot 9 \cdot 4$

$V = 12\pi$

20. $S = 4\pi r^2$

$S = 4\pi r(2)^2$

$S = 4\pi \cdot 4$

$S = 16\pi$

21. $2x = 14$

$x = 7$

22. $3x = -3$

$x = -1$

23. $\frac{y}{-3} = 4$

$y = -12$

24. $7y = -21$

$y = -3$

25. $2a + 2 = 26$

$2a = 24$

$a = 12$

26. $\frac{3b}{2} = 27$

$b = 18$

27. $2x + 2 = 30 - 6x + 12$

$2x + 2 = 42 - 6x$

$8x = 40$

$x = 5$

28. $2x + 2 + 3x + 6 = 22 + 40 - 4x$

$5x + 8 = 62 - 4x$

$9x = 54$

$x = 6$

29. Multiply equation by 6 to get

$2x - 3x = -30$

$-x = -30$

$x = 30$

30. Multiply equation by 12 to get

$6x + 4x + 3x = 312$

$13x = 312$

$x = 24$

31. Multiply equation by n to get

$360 + 135n = 180n$

$360 = 45n$

$8 = n$

32. Multiply equation by n to get

$(n - 2)180 = 150n$

$180n - 360 = 150n$

$-360 = -30n$

$12 = n$

33. $148 = 2 \cdot 5 \cdot w + 2 \cdot w \cdot 6 + 2 \cdot 5 \cdot 6$

$148 = 10w + 12w + 60$

$88 = 22w$

$4 = w$

34. $156 = \left(\frac{1}{2}\right) \cdot 12 \cdot (b + 11)$

$156 = 6(b + 11)$

$156 = 6b + 66$

$90 = 6b$

$15 = b$

35. $23 = \left(\frac{1}{2}\right)(78 - y)$

$46 = 78 - y$

$-32 = -y$

$32 = y$

36. $\frac{-3}{2} = \frac{Y - 1}{2 - (-2)}$

$\frac{-3}{2} = \frac{Y - 1}{4}$

$2Y - 2 = -12$

$2Y = -10$

$Y = -5$

37. Since $a \cdot c$ is the product of two real numbers, it is a real number, too.

$\therefore\ a \cdot c = a \cdot c$ (Reflexive Axiom)

But $a = b$ (Hypothesis)

$\therefore\ a \cdot c = b \cdot c$ (Substitution)

38. Because $a - c$ is the difference of two real numbers, it is a real number, too.

$\therefore\ a - c = a - c$ (Reflexive Axiom)

But $a = b$ (Hypothesis)

$\therefore\ a - c = b - c$ (Substitution)

39. Let $a = 2$, $b = 3$, $c = 0$;

$2 \cdot 0 = 3 \cdot 0$ but $2 \neq 3$.

40. Let $a = 3$ and $b = -3$;

$(3)^2 = (-3)^2$ but $3 \neq -3$.

A.3: Inequalities

1. The length of $\overline{AB}$ is greater than the length of $\overline{CD}$.
2. $e < f$; $f > e$
3. The measure of angle ABC is greater than the measure of angle DEF.
4. $x = 6$, $x = 9$, $x = 12$
5. a. $p = 4$

   b. $p = 10$
6. Yes
7. AB > IJ
8. The measure of angle JKL is greater than the measure of angle ABC.
9. a. False

   b. True

   c. True

   d. False
10. a. True

    b. False

    c. True

    d. False
11. The measure of the second angle must be greater than 148 and less than 180.
12. The length of the second board must be less than 5 feet.
13. a. $-12 \leq 20$

    b. $-10 \leq -2$

    c. $18 \geq -30$

    d. $3 \geq -5$
14. a. $2 > -1$

    b. $12 < 18$

    c. $-12 > -18$

    d. $2 < 3$
15.

No Change	No Change
No Change	No Change
No Change	CHANGE
No Change	CHANGE

16. $5x \leq 30$

$x \leq 6$

17. $2x \leq 14$

$x \leq 7$

18. $4x > 20$

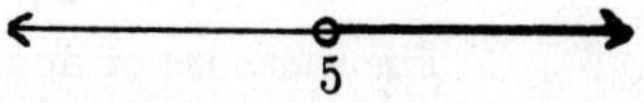

$x > 5$

19. $-4x > 20$

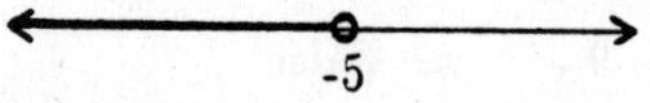

$x < -5$

20. $10 - 5x \leq 30$

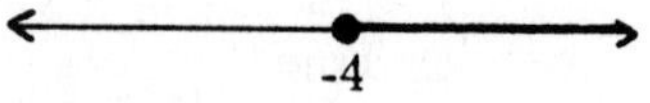

$-5x \leq 20$

$x \geq -4$

21. $5x < 200 - 5x$

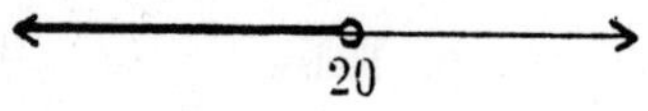

$10x < 200$

$x < 20$

22. $5x + 10 < 54 - 6x$

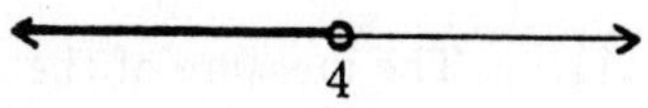

$11x < 44$

$x < 4$

23. $2x - 3x \leq 24$

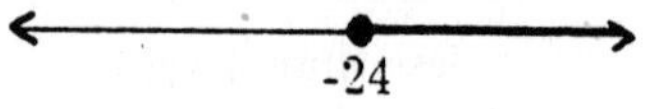

$-x \leq 24$

$x \geq -24$

24. $2x - 3 < -35$

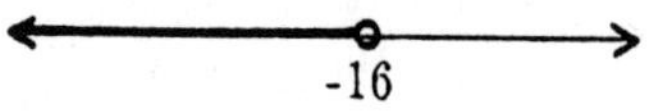

$2x < -32$

$x < -16$

25. $x^2 + 4x \leq x^2 - 5x - 18$

$9x \leq -18$

$x \leq -2$

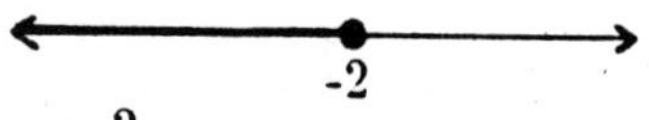

26. $x^2 + 2x < 2x - x^2 + 2x^2$

$x^2 + 2x < x^2 + 2x$

No solution or $\emptyset$

27. Not true if $c < 0$.

28. Not true if $c = 0$.

29. Not true if $a = -3$ and $b = -2$.

30. Not true if $a = c$.

31. If $a < b$ and $c < d$, then $a + p_1 = b$ and $c + p_2 = d$ where p_1 and p_2 are both positive. Use Addition Property to get

$a + p_1 + c + p_2 = b + d$

$a + c + p_1 + p_2 = b + d$

But since p_1 and p_2 are both positive, $p_1 + p_2$ must also be positive.

$\therefore a + c < b + d$.

32. If $a < b$, then $a + p_1 = b$ where p_1 is positive. Multiply by -1.

$-a - p_1 = -b$.

Add p_1 to both sides

$-a = -b + p_1$

Add c to both sides

$c + (-a) = c + (-b) + p_1$

$c - a = c - b + p_1$

Since p_1 is positive, $c - a > c - b$.

A.4: Quadratic Equations

1. a. 3.61

   b. 2.83

   c. -5.39

   d. 0.77

2. a. 4.12

   b. 20

   c. -2.65

   d. 1.26

3. a, c, d, f

4. a, b, c, e

5. a. $\sqrt{8} = \sqrt{4 \cdot 2} = 2\sqrt{2}$

   b. $\sqrt{45} = \sqrt{9 \cdot 5} = 3\sqrt{5}$

   c. $\sqrt{900} = 30$

   d. $(\sqrt{3})^2 = 3$

6. a. $\sqrt{28} = \sqrt{4 \cdot 7} = 2\sqrt{7}$

   b. $\sqrt{32} = \sqrt{16 \cdot 2} = 4\sqrt{2}$

   c. $\sqrt{54} = \sqrt{9 \cdot 6} = 3\sqrt{6}$

   d. $\sqrt{200} = \sqrt{100 \cdot 2} = 10\sqrt{2}$

7. a. $\sqrt{\frac{9}{16}} = \frac{\sqrt{9}}{\sqrt{16}} = \frac{3}{4}$

   b. $\sqrt{\frac{25}{49}} = \frac{\sqrt{25}}{\sqrt{49}} = \frac{5}{7}$

   c. $\sqrt{\frac{7}{16}} = \frac{\sqrt{7}}{\sqrt{16}} = \frac{\sqrt{7}}{4}$

   d. $\sqrt{\frac{6}{9}} = \frac{\sqrt{6}}{\sqrt{9}} = \frac{\sqrt{6}}{3}$

8. a. $\sqrt{\frac{1}{4}} = \frac{\sqrt{1}}{\sqrt{4}} = \frac{1}{2}$

   b. $\sqrt{\frac{16}{9}} = \frac{\sqrt{16}}{\sqrt{9}} = \frac{4}{3}$

   c. $\sqrt{\frac{5}{36}} = \frac{\sqrt{5}}{\sqrt{36}} = \frac{\sqrt{5}}{6}$

   d. $\sqrt{\frac{3}{16}} = \frac{\sqrt{3}}{\sqrt{16}} = \frac{\sqrt{3}}{4}$

9. a. $\sqrt{54} \approx 7.35$ and $3\sqrt{6} \approx 7.35$

   b. $\sqrt{\frac{5}{16}} \approx 0.56$ and $\frac{\sqrt{5}}{4} \approx 0.56$

10. a. $\sqrt{48} \approx 6.93$ and $4\sqrt{3} \approx 6.93$

    b. $\sqrt{\frac{7}{9}} \approx 0.88$ and $\frac{\sqrt{7}}{3} \approx 0.88$

11. $x^2 - 6x + 8 = 0$

    $(x - 4)(x - 2) = 0$

    $x - 4 = 0$ or $x - 2 = 0$

    $x = 4$ or $x = 2$

12. $x^2 + 4x = 21$

    $x^2 + 4x - 21 = 0$

    $(x + 7)(x - 3) = 0$

    $x + 7 = 0$ or $x - 3 = 0$

    $x = -7$ or $x = 3$

13. $3x^2 - 51x + 180 = 0$

    $3(x^2 - 17x + 60) = 0$

    $3(x - 12)(x - 5) = 0$

    $x - 12 = 0$ or $x - 5 = 0$

    $x = 12$ or $x = 5$

14. $2x^2 + x - 6 = 0$

    $(2x - 3)(x + 2) = 0$

    $2x - 3 = 0$ or $x + 2 = 0$

    $2x = 3$ or $x = -2$

    $x = \frac{3}{2}$ or $x = -2$

15. $3x^2 = 10x + 8$

    $3x^2 - 10x - 8 = 0$

    $(3x + 2)(x - 4) = 0$

    $3x + 2 = 0$ or $x - 4 = 0$

    $3x = -2$ or $x = 4$

    $x = -\frac{2}{3}$ or $x = 4$

16. $8x^2 + 40x - 112 = 0$

    $8(x^2 + 5x - 14) = 0$

    $8(x + 7)(x - 2) = 0$

    $x + 7 = 0$ or $x - 2 = 0$

    $x = -7$ or $x = 2$

17. $6x^2 = 5x - 1$

$6x^2 - 5x + 1 = 0$

$(3x - 1)(2x - 1) = 0$

$3x - 1 = 0$ or $2x - 1 = 0$

$3x = 1$ or $2x = 1$

$x = \frac{1}{3}$ or $x = \frac{1}{2}$

18. $12x^2 + 10x = 12$

$12x^2 + 10x - 12 = 0$

$2(6x^2 + 5x - 6) = 0$

$2(3x - 2)(2x + 3) = 0$

$3x - 2 = 0$ or $2x + 3 = 0$

$3x = 2$ or $2x = -3$

$x = \frac{2}{3}$ or $x = -\frac{3}{2}$

19. $x^2 - 7x + 10 = 0$

$a = 1 \quad b = -7 \quad c = 10$

$$x = \frac{-b \pm \sqrt{b^2 - 4ac}}{2a}$$

$$x = \frac{7 \pm \sqrt{49 - 4(1)(10)}}{2(1)}$$

$$x = \frac{7 \pm \sqrt{49 - 40}}{2}$$

$$x = \frac{7 \pm \sqrt{9}}{2}$$

$$x = \frac{7 + 3}{2} \text{ or } x = \frac{7 - 3}{2}$$

$x = 5$ or 2

20. $x^2 + 7x + 12 = 0$

$a = 1 \quad b = 7 \quad c = 12$

$$x = \frac{-b \pm \sqrt{b^2 - 4ac}}{2a}$$

$$x = \frac{-7 \pm \sqrt{49 - 4(1)(12)}}{2(1)}$$

$$x = \frac{-7 \pm \sqrt{49 - 48}}{2}$$

$$x = \frac{-7 \pm \sqrt{1}}{2}$$

$$x = \frac{-7 + 1}{2} \text{ or } \frac{-7 - 1}{2}$$

$x = -3$ or -4

21. $x^2 + 9 = 7x$

$x^2 - 7x + 9 = 0$

$a = 1 \quad b = -7 \quad c = 9$

$$x = \frac{-b \pm \sqrt{b^2 - 4ac}}{2a}$$

$$x = \frac{7 \pm \sqrt{49 - 4(1)(9)}}{2(1)}$$

$$x = \frac{7 \pm \sqrt{49 - 36}}{2}$$

$$x = \frac{7 \pm \sqrt{13}}{2} \approx 5.30 \text{ or } 1.70$$

22. $2x^2 + 3x = 6$

$2x^2 + 3x - 6 = 0$

$a = 2 \quad b = 3 \quad c = -6$

$$x = \frac{-b \pm \sqrt{b^2 - 4ac}}{2a}$$

$$x = \frac{-3 \pm \sqrt{9 - 4(2)(-6)}}{2(2)}$$

$$x = \frac{-3 \pm \sqrt{9 + 48}}{4}$$

$$x = \frac{-3 \pm \sqrt{57}}{4} \approx 1.14 \text{ or } -2.64$$

23. $x^2 - 4x - 8 = 0$

$a = 1 \quad b = -4 \quad c = -8$

$$x = \frac{-b \pm \sqrt{b^2 - 4ac}}{2a}$$

$$x = \frac{4 \pm \sqrt{16 - 4(1)(-8)}}{2(1)}$$

$$x = \frac{4 \pm \sqrt{16 + 32}}{2}$$

$$x = \frac{4 \pm \sqrt{48}}{2}$$

$$x = \frac{4 \pm \sqrt{16 \cdot 3}}{2}$$

$$x = \frac{4 \pm 4\sqrt{3}}{2}$$

$x = 2 \pm 2\sqrt{3} \approx 5.46$ or -1.46

24. $x^2 - 6x - 2 = 0$

$a = 1 \quad b = -6 \quad c = -2$

$$x = \frac{-b \pm \sqrt{b^2 - 4ac}}{2a}$$

$$x = \frac{6 \pm \sqrt{36 - 4(1)(-2)}}{2(1)}$$

$$x = \frac{6 \pm \sqrt{36 + 8}}{2}$$

$$x = \frac{6 \pm \sqrt{44}}{2}$$

$$x = \frac{6 \pm \sqrt{4 \cdot 11}}{2}$$

$$x = \frac{6 \pm 2\sqrt{11}}{2}$$

$x = 3 \pm \sqrt{11} \approx 6.32$ or -0.32

25. $5x^2 = 3x + 7$

$5x^2 - 3x - 7 = 0$

$a = 5 \quad b = -3 \quad c = -7$

$$x = \frac{-b \pm \sqrt{b^2 - 4ac}}{2a}$$

$$x = \frac{3 \pm \sqrt{9 - 4(5)(-7)}}{2(5)}$$

$$x = \frac{3 \pm \sqrt{9 + 140}}{10}$$

$x = \dfrac{3 \pm \sqrt{149}}{10} \approx 1.52$ or -0.92

26. $2x^2 = 8x - 1$

$2x^2 - 8x + 1 = 0$

$a = 2 \quad b = -8 \quad c = 1$

$$x = \frac{-b \pm \sqrt{b^2 - 4ac}}{2a}$$

$$x = \frac{8 \pm \sqrt{64 - 4(2)(1)}}{2(2)}$$

$$x = \frac{8 \pm \sqrt{64 - 8}}{4}$$

$$x = \frac{8 \pm \sqrt{56}}{4}$$

$$x = \frac{8 \pm \sqrt{4 \cdot 14}}{4}$$

$$x = \frac{8 \pm 2\sqrt{14}}{4}$$

$x = \dfrac{4 \pm \sqrt{14}}{2} \approx 3.87$ or 0.13

27. $2x^2 = 14$

$x^2 = 7$

$x = \pm\sqrt{7}$

$x \approx \pm 2.65$

28. $2x^2 = 14x$

$2x^2 - 14x = 0$

$2x(x - 7) = 0$

$2x = 0$ or $x - 7 = 0$

$x = 0$ or $x = 7$

29. $4x^2 - 25 = 0$

$4x^2 = 25$

$x^2 = \frac{25}{4}$

$x = \pm\frac{5}{2}$

30. $4x^2 - 25x = 0$

$x(4x - 25) = 0$

$x = 0$ or $4x - 25 = 0$

$x = 0$ or $4x = 25$

$x = 0$ or $x = \frac{25}{4}$

31. $ax^2 - bx = 0$

$x(ax - b) = 0$

$x = 0$ or $ax - b = 0$

$x = 0$ or $ax = b$

$x = 0$ or $x = \frac{b}{a}$

32. $ax^2 - b = 0$

$ax^2 = b$

$x^2 = \frac{b}{a}$

$x = \pm\sqrt{\frac{b}{a}}$

$x = \pm\frac{\sqrt{ab}}{a}$

33. Let the length $= x + 3$ and width $= x$. The area is then:

$x(x + 3) = 40$

$x^2 + 3x = 40$

$x^2 + 3x - 40 = 0$

$(x + 8)(x - 5) = 0$

$x + 8 = 0$ or $x - 5 = 0$

$x = -8$ or $x = 5$

Reject $x = -8$ because the length cannot be negative.

The rectangle is 5 by 8.

34. $x \cdot (x + 5) = (x + 1) \cdot 4$

$x^2 + 5x = 4x + 4$

$x^2 + 1x - 4 = 0$

$a = 1 \quad b = 1 \quad c = -4$

$$x = \frac{-b \pm \sqrt{b^2 - 4ac}}{2a}$$

$$x = \frac{-1 \pm \sqrt{1 - 4(1)(-4)}}{2(1)}$$

$$x = \frac{-1 \pm \sqrt{1 + 16}}{2}$$

$$x = \frac{-1 \pm \sqrt{17}}{2}$$

$$CP = \frac{-1 + \sqrt{17}}{2} \approx 1.56;$$

$\frac{-1 - \sqrt{17}}{2}$ is rejected because it is a negative number.

35. $D = \frac{n(n - 3)}{2}$

$9 = \frac{n(n - 3)}{2}$

$18 = n^2 - 3n$

$0 = n^2 - 3n - 18$

$0 = (n - 6)(n + 3)$

$n - 6 = 0$ or $n + 3 = 0$

$n = 6$ or $n = -3$

$n = 6$; reject $n = -3$

36. $D = \frac{n(n - 3)}{2}$

$n = \frac{n(n - 3)}{2}$

$2n = n^2 - 3n$

$0 = n^2 - 5n$

$0 = n(n - 5)$

$n = 0$ or $n - 5 = 0$

$n = 0$ or $n = 5$

$n = 5$; reject $n = 0$

37. $c^2 = a^2 + b^2$

$c^2 = 3^2 + 4^2$

$c^2 = 9 + 16$

$c^2 = 25$

$c = \pm 5$

$c = 5$; reject $c = -5$

38. $c^2 = a^2 + b^2$

$10^2 = 6^2 + b^2$

$100 - 36 = b^2$

$64 = b^2$

$b = \pm 8$

$b = 8$; reject $b = -8$

Appendix B

B.1: Truth Tables

1. $F \vee F = F$
2. $T \vee T = T$
3. $T \wedge T = T$
4. $T \wedge F = F$
5. $T \rightarrow F = F$
6. $T \rightarrow T = T$
7. $T \wedge (T \vee F) = T \wedge T = T$
8. $(T \wedge T) \vee F = T \vee F = T$
9. Hamburgers are not health food.
10. Mary is an accouontant or hamburgers are health food.
11. If Mary is an accountant, then hamburgers are health food.
12. Mary is an accountant and hamburgers are not health food.

13.

P	~P	P ∨ ~P
T	F	T
F	T	T

Yes

14.

P	~P	P ∧ ~P
T	F	F
F	T	F

No

15.

P	Q	(P ∨ Q)	(P ∨ Q) → P
T	T	T	T
T	F	T	T
F	T	T	F
F	F	F	T

No

16.

P	Q	(P ∧ Q)	(P ∧ Q) → Q
T	T	T	T
T	F	F	T
F	T	F	T
F	F	F	T

Yes

17.

P	Q	(P → Q)	(P → Q) ∧ Q)	[(P→Q) ∧ Q)] → P
T	T	T	T	T
T	F	F	F	T
F	T	T	T	F
F	F	T	F	T

No

18.

P	Q	(P → Q)	(P → Q) ∧ P)	[(P→Q) ∧ P)] → Q
T	T	T	T	T
T	F	F	F	T
F	T	T	F	T
F	F	T	F	T

Yes

19. $\sim(P \land Q) \to \sim P \lor \sim Q$

20. $\sim(P \lor Q) \to \sim P \land \sim Q$

21. Mary is not an accountant and hamburgers are not health food.

22. Mary is not an accountant or hamburgers are not health food.

23. It is not cold or it is not snowing.

24. We will not go to dinner and we will not go the the movie.

25. Prove: $[\sim(P \vee Q)] \leftrightarrow [\sim P \wedge \sim Q]$

P	Q	$(P \vee Q)$	$[\sim(P \vee Q)]$	$\sim P$	$\sim Q$	$[\sim P \wedge \sim Q]$
T	T	T	F	F	F	F
T	F	T	F	F	T	F
F	T	T	F	T	F	F
F	F	F	T	T	T	T

The columns for $[\sim(P \vee Q)]$ and $[\sim P \wedge \sim Q]$ are the same.

26. Prove: $[Q \rightarrow P] \leftrightarrow [\sim P \rightarrow \sim Q]$

P	Q	$[Q \rightarrow P]$	$\sim P$	$\sim Q$	$[\sim P \rightarrow \sim Q]$
T	T	T	F	F	T
T	F	T	F	T	T
F	T	F	T	F	F
F	F	T	T	T	T

The columns for $[Q \rightarrow P]$ and $[\sim P \rightarrow \sim Q]$ are the same.

27. Show $[(P \rightarrow Q) \wedge (Q \rightarrow R)] \rightarrow (P \rightarrow R)$ is a tautology.

P	Q	R	$(P \rightarrow Q)$	$(Q \rightarrow R)$	$[(P \rightarrow Q) \wedge (Q \rightarrow R)]$	$(P \rightarrow R)$	$[(P \rightarrow Q) \wedge (Q \rightarrow R)] \rightarrow (P \rightarrow R)$
T	T	T	T	T	T	T	T
T	T	F	T	F	F	F	T
T	F	T	F	T	F	T	T
T	F	F	F	T	F	F	T
F	T	T	T	T	T	T	T
F	T	F	T	F	F	T	T
F	F	T	T	T	T	T	T
F	F	F	T	T	T	T	T

28. Show $[P \vee (Q \wedge R)] \leftrightarrow [(P \vee Q) \wedge (P \vee R)]$

P	Q	R	$(Q \wedge R)$	$[P \vee (Q \wedge R)]$	$(P \vee Q)$	$(P \vee R)$	$[(P \vee Q) \wedge (P \vee R)]$
T	T	T	T	T	T	T	T
T	T	F	F	T	T	T	T
T	F	T	F	T	T	T	T
T	F	F	F	T	T	T	T
F	T	T	T	T	T	T	T
F	T	F	F	F	T	F	F
F	F	T	F	F	F	T	F
F	F	F	F	F	F	F	F

The columns for $[P \vee (Q \wedge R)]$ and $[(P \vee Q) \wedge (P \vee R)]$ are the same.

29. Show $[P \wedge \sim Q]$ is the negative of $P \rightarrow Q$.

P	Q	$\sim Q$	$[P \wedge \sim Q]$	$(P \rightarrow Q)$
T	T	F	F	T
T	F	T	T	F
F	T	F	F	T
F	F	T	F	T

The columns for $[P \wedge \sim Q]$ and $P \rightarrow Q$ are opposites.

30. It is medicine and it does not taste bad.

31. I am good and I cannot go to the movie.

32. I am 18 or older and I cannot vote.

33. I studied hard and made an A and I cannot be a member of Phi Theta Kappa.

1 . The sum of the measures of $\angle$s 1 and 2 is 90 degrees.

2 . We will have to turn on the air conditioner.

3 . Tina will have a good time.

4 . We will go to the concert.

5 . $\angle 1$ and $\angle 2$ are not complementary.

6 . Fido does not live in the zoo.

7 . Tom finished the job.

8 . The traffic light did not change.

9 . If Izzi lives in Chicago, then she lives in the Midwest.

10. If you pay your tuition, then you will need to write a check.

11. If Ken Travis gets a hit, then I will be happy.

12. If Tom speaks at the rally, the union will decide not to strike.

13. Valid

14. Not valid

15. Not valid

16. Valid

17. Law of Syllogism

18. Law of Negative Inference

19. Law of Detachment

20. Law of Syllogism

21. a. $[(P \vee Q) \wedge \sim Q] \rightarrow P$

b.

P	Q	$(P \vee Q)$	$\sim Q$	$(P \vee Q) \wedge \sim Q$	$[(P \vee Q) \wedge \sim Q] \rightarrow P$
T	T	T	F	F	T
T	F	T	T	T	T
F	T	T	F	F	T
F	F	F	T	F	T

22. Terry is sick.

23. Mary's family will visit at Christmas.

24. Wendell had to study geometry.

25. Law of Detachment: $[(P \rightarrow Q) \land P] \rightarrow Q$

P	Q	(P → Q)	(P → Q) ∧ P	[(P → Q) ∧ P] → Q
T	T	T	T	T
T	F	F	F	T
F	T	T	F	T
F	F	T	F	T

26. Law of Negative Inference: $[(P \rightarrow Q) \land \sim Q] \rightarrow \sim P$

P	Q	(P → Q)	~Q	[(P → Q) ∧ ~Q]	~P	[(P → Q) ∧ ~Q] → ~P
T	T	T	F	F	F	T
T	F	F	T	F	F	T
F	T	T	F	F	T	T
F	F	T	T	T	T	T

27. Law of Syllogism: $[(P \rightarrow Q) \land (Q \rightarrow R)] \rightarrow (P \rightarrow R)$

P	Q	R	(P → Q)	(Q → R)	[(P → Q) ∧ (Q → R)]	(P → R)	[(P → Q) ∧ (Q → R)] → (P → R)
T	T	T	T	T	T	T	T
T	T	F	T	F	F	F	T
T	F	T	F	T	F	T	T
T	F	F	F	T	F	F	T
F	T	T	T	T	T	T	T
F	T	F	T	F	F	T	T
F	F	T	T	T	T	T	T
F	F	F	T	T	T	T	T